Status of Recent Geoscience Graduates 2014

Carolyn Wilson
American Geosciences Institute
Alexandria, VA 22302

About AGI's Geoscience Student Exit Survey

The American Geosciences Institute (AGI) launched the Geoscience Student Exit Survey to assist geoscience departments in assessing the educational experiences of graduating students, as well as to examine ongoing evidence of knowledge gaps of new graduates entering the workforce. With this survey we hope to identify student decision points for entering and persisting within the geosciences, identify the geoscience research fields and co-curricular activities of interest to the students, identify the preferred jobs and industries of graduating students including those not considered part of the traditional geoscience workforce, and establish a benchmark for a detailed study of the career pathways of early career geoscientists. Likewise, the Geoscience Student Exit Survey is the gateway for an ongoing longitudinal survey of geoscience professionals that was launched in 2013.

The report examines the responses to AGI's Geoscience Student Exit Survey by graduates from the 2013-2014 academic year.

This survey has four major sections: student demographics, educational background, postsecondary education experiences, and post-graduation plans, with specific questions that cover areas such as community college experiences, quantitative skills, field and research experiences, internships, and details about their immediate plans for graduate school or in a new job. The survey was piloted twice in spring 2011 and spring 2012. For spring 2013 graduation, we opened the survey up to all geoscience department in the United States, and they were asked to pass along the survey link for their department to their graduating students. As an incentive for participating, each department has the opportunity to add questions to the survey for their particular graduates to answer. At the end of the survey period, all the data from students from a particular department is shared in aggregate with that department as long as they have a sufficient number of participating students to ensure individual response privacy. In 2014, the survey was available for graduates at the end of each semester — fall, spring, and summer.

AGI is working to expand the distribution of the survey starting with geoscience departments in the United Kingdom. As we move forward, we hope to keep recruiting more countries to participate in AGI's Geoscience Student Exit Survey. This survey will provide the global geosciences community with a more detailed description of the make-up and preparation of the early career geoscientist as he/she enters the workforce.

If you would like more information or would like your department to participate in AGI's Geoscience Student Exit Survey, please contact Carolyn Wilson.

Status of Recent Geoscience Graduates 2014

Edited by Carolyn Wilson

ISBN: 0-922152-99-3
ISBN-13: 978-0-922152-99-5

Graphs by Carolyn Wilson, AGI Workforce Program
Design by Brenna Tobler, AGI Graphic Designer

For more information on the American Geosciences Institute and its publications check us out at www.americangeosciences.org/pubs.

© 2014 American Geosciences Institute.

AGI american geosciences institute

connecting earth, science, and people

Carolyn Wilson, Geoscience Workforce Data Analyst
American Geosciences Institute
4220 King Street, Alexandria, VA 22302
www.americangeosciences.org
cwilson@americangeosciences.org
(703) 379-2480, ext. 632

Front cover photo © Alison Dorsey, back cover photo © Matt Smith. All photos in this report were submitted to the 2014 Life in the Field contest, which requested images representing meaningful geoscience work through internships, research, employment, or field experiences.

GEOSCIENCE STUDENT EXIT SURVEY

Executive Summary

The American Geosciences Institute's (AGI) Status of Recent Geoscience Graduates 2014 provides an overview of the demographics, activities, and experiences of geoscience students that received their bachelor's, master's, or doctoral degree during the 2013-2014 academic year. This research draws attention to student preparation in the geosciences and their education and career path decisions, as well as examines many of the questions raised about student transitions into the workforce.

The Status of Recent Geoscience Graduates report was first released in 2013 presenting data from spring 2013 graduates. For the 2014 report, the number of participants in the AGI's Geoscience Student Exit Survey increased by 60% compared to 2013 creating a sample size that better represents the community of geoscience graduates. However, many of the trends seen in 2013 are echoed in the 2014 report.

This report presents the results for the end user's consideration. Two notable trends seen in this report are related to the quantitative skills and knowledge of the graduates and their plans for the immediate future. As in 2013, most geoscience graduates complete Calculus II during their postsecondary education, but the rate of graduates taking higher-level quantitative coursework drops quickly. However, in 2014, there was an overall decrease in the rate of graduates that took Statistics, even though an understanding of statistics is often necessary for graduate research. In addition, the type of school the students attended may affect their access to some of these higher-level quantitative courses beyond Calculus I.

This year also saw an increase in the rate of bachelor's and master's graduates planning to attend graduate school and/or still seeking employment in the geosciences. While jobs are currently available in the geosciences, many graduates appear to have struggled finding employment, which raises the concern about the preparedness of these graduates for entering the geoscience workforce upon graduation. When asked why they were looking for employment outside of the geosciences, the graduates mentioned trouble finding employers at geoscience industries willing to hire them with their little to no work experience, but the majority of graduates do not participate in an internship during their postsecondary education. Also, based on the graduation rates collected by AGI over the years, most of the graduate programs across the country are near capacity, which could increase the difficulty for these recent graduates to secure a position within these programs.

As participation continues to grow, we expect to provide more details and comparisons regionally and by degree levels. AGI is also conducting an ongoing longitudinal survey that will follow these early-career geoscientists as they continue to progress in the workforce. The combination of these two studies will present a better understanding of the workforce pathway in the geosciences and enable both geoscience departments and employers to improve the educational and career opportunities for future geoscience graduates.

Acknowledgements

I would like to recognize a few organization and individuals for their support for this project. Thanks to ConocoPhillips for their support towards the project this year. Thanks also to the American Institute of Professional Geologists and the Geological Society of America for distributing the survey to their student membership. I also want to thank the AGI Workforce 2014 Summer Intern, Sebastian Corrochano, for his hard work quality controlling and organizing the data from this survey, and the AGI Workforce 2014 Fall Intern, Jamie Ricci, for her help coding the qualitative responses. Finally, I would especially like to thank the department contacts from each participating department for distributing the survey to their graduating students.

Contents

An Overview of the Demographics of the Participants

This year, AGI's Geoscience Student Exit Survey was made available to geoscience graduates at all traditional graduation periods (winter, spring, and summer) during the 2013-2014 academic year, to be collectively referenced as from 2014. Approximately two months before the end of each semester, an email was sent to all the heads and chairs of geoscience departments asking for their participation in this study. As an incentive to participate, AGI gives the departments the data in aggregate for their graduates for their internal assessment purposes. Distribution instructions and the survey link were sent to the identified representatives for each department that agreed to send the survey to their graduates. Departments continue to have the option to customize the survey appropriately for their graduates.

This year, AGI also asked the American Institute for Professional Geologists (AIPG) and the Geological Society of America (GSA) to distribute the survey link to their student membership, which helped increase participation dramatically. In the future, AGI hopes to recruit more member societies to help with distribution of the survey.

The survey was available to winter and summer graduates for two months, and spring graduates had three months to complete the survey. At the close of the survey, 688 graduating students from 163 geoscience schools or departments provided responses — 517 bachelor's graduates, 115 master's graduates, and 56 doctoral graduates. This is a 60 percent increase in participation from last year, and using AGI's graduation data from 2013, this sample size was determined to be sufficiently large enough to statistically represent the total population of geoscience graduates.

This first section of the survey covered student demographics to establish an understanding of the students that graduate in the geosciences. Even with the increased participation, the data are similar to the data collected in 2013. However, there is a shift in the gender dynamics. The percentage of female master's graduates increased to nearly equal to male master's graduates, and in 2014, the percentage of female doctoral graduates exceeded the percentage of male doctoral graduates by 11%. As in 2013, students indicating their citizenship as U.S. citizen or permanent resident were asked to indicate their race and ethnicity. The percentage of underrepresented minorities include African Americans, Hispanic/Latinos, Native Americans/Alaskans, and Native Hawaiians/Pacific Islanders. However, it is important to note that the percentage of underrepresented minorities is dominated by the Hispanic/Latino population of geoscience graduates. For the 2014 report, a figure showing the distribution of graduates by their age was added.

Photo by Arnaud Mansat, Roman Teisserenc, and Allison Myers-Pigg from the AGI 2014 Life in the Field photo contest.

Distribution of participating graduating students and departments*

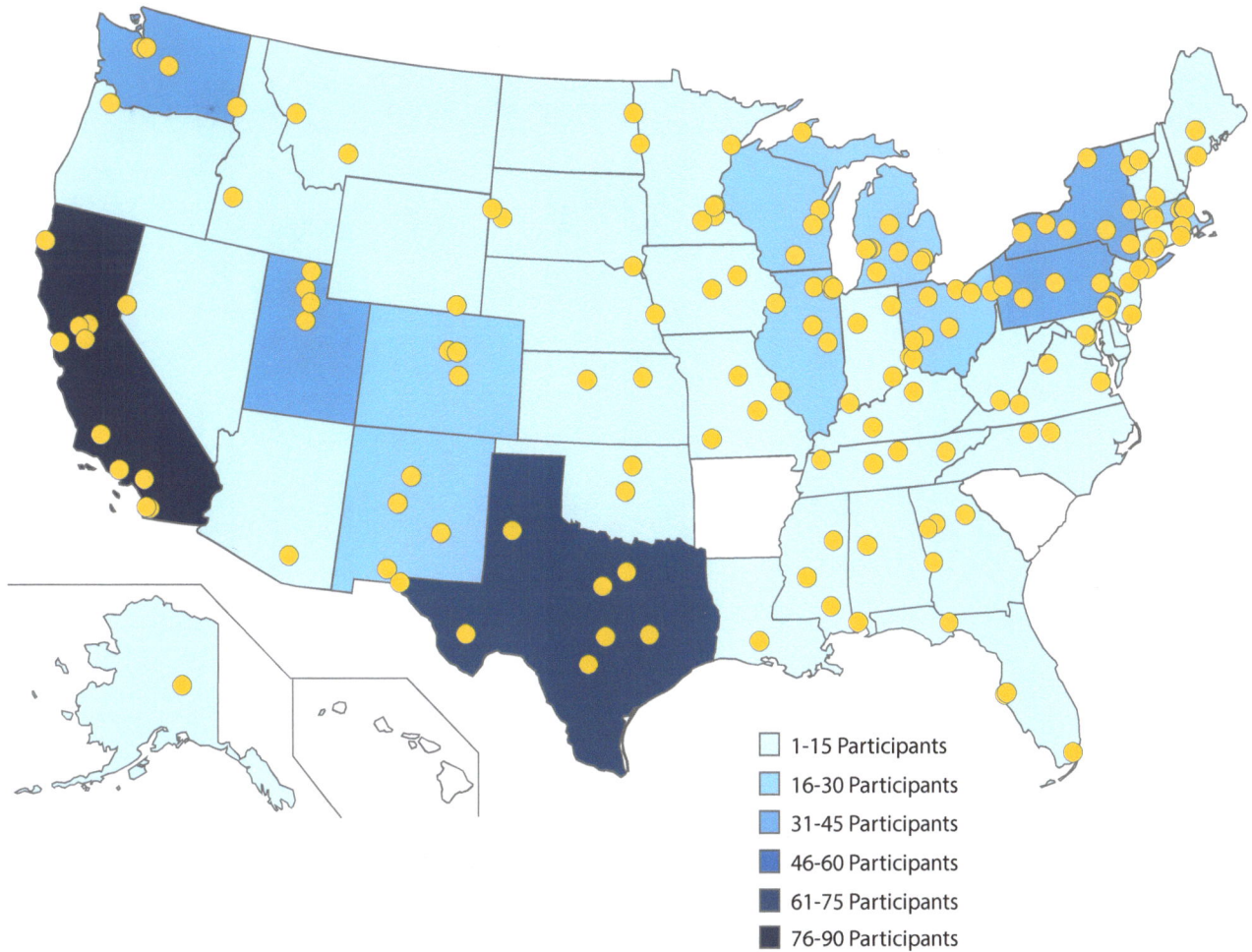

Legend:
- 1-15 Participants
- 16-30 Participants
- 31-45 Participants
- 46-60 Participants
- 61-75 Participants
- 76-90 Participants

The relative distribution by state of the universities and their graduating geoscience students across the United States that participated in the Exit Survey. *See Appendix I for list of departments

Degree received by participating graduates

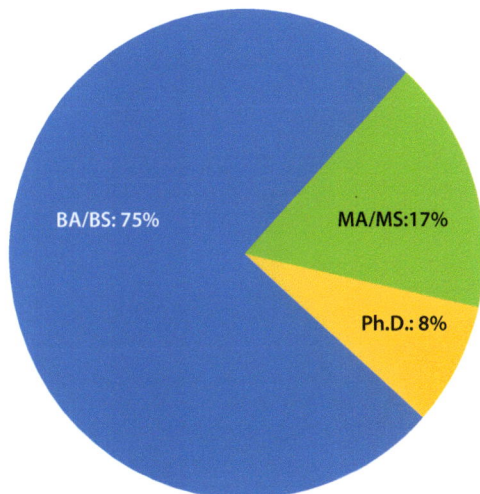

BA/BS: 75%
MA/MS: 17%
Ph.D.: 8%

Percentage of respondents within different classified institutions**

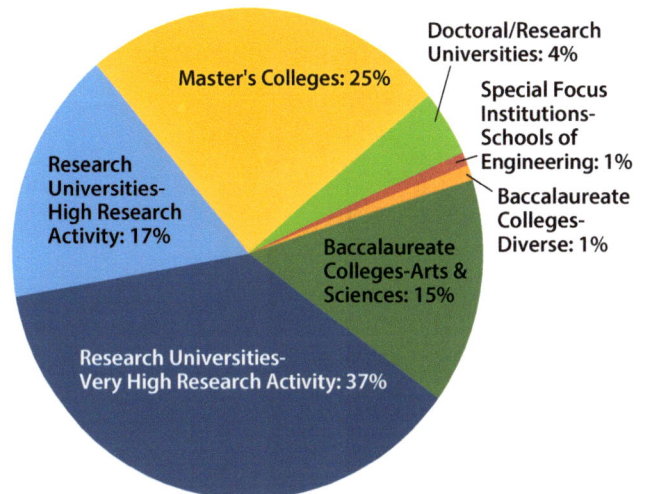

Master's Colleges: 25%

Doctoral/Research Universities: 4%

Special Focus Institutions-Schools of Engineering: 1%

Baccalaureate Colleges-Diverse: 1%

Research Universities-High Research Activity: 17%

Baccalaureate Colleges-Arts & Sciences: 15%

Research Universities-Very High Research Activity: 37%

**See Appendix II for definitions of the Carnegie University Classification System

Gender breakdown of graduates

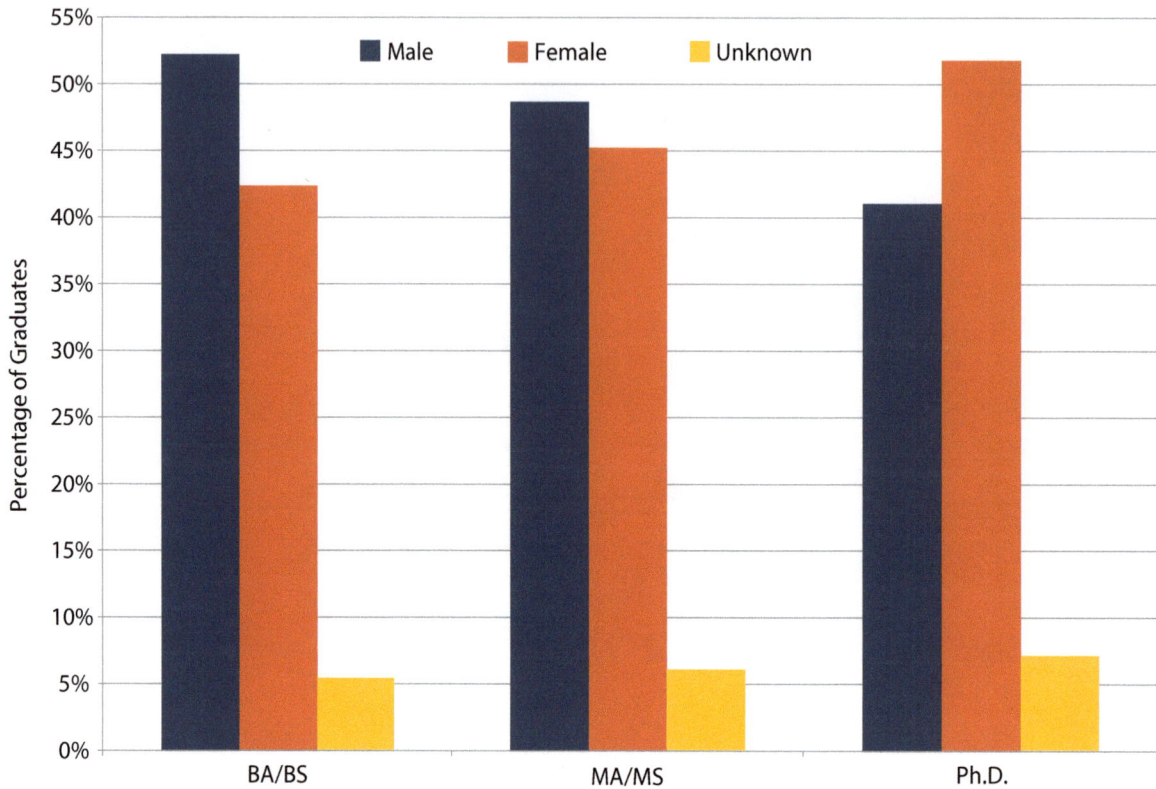

Age distribution of graduates

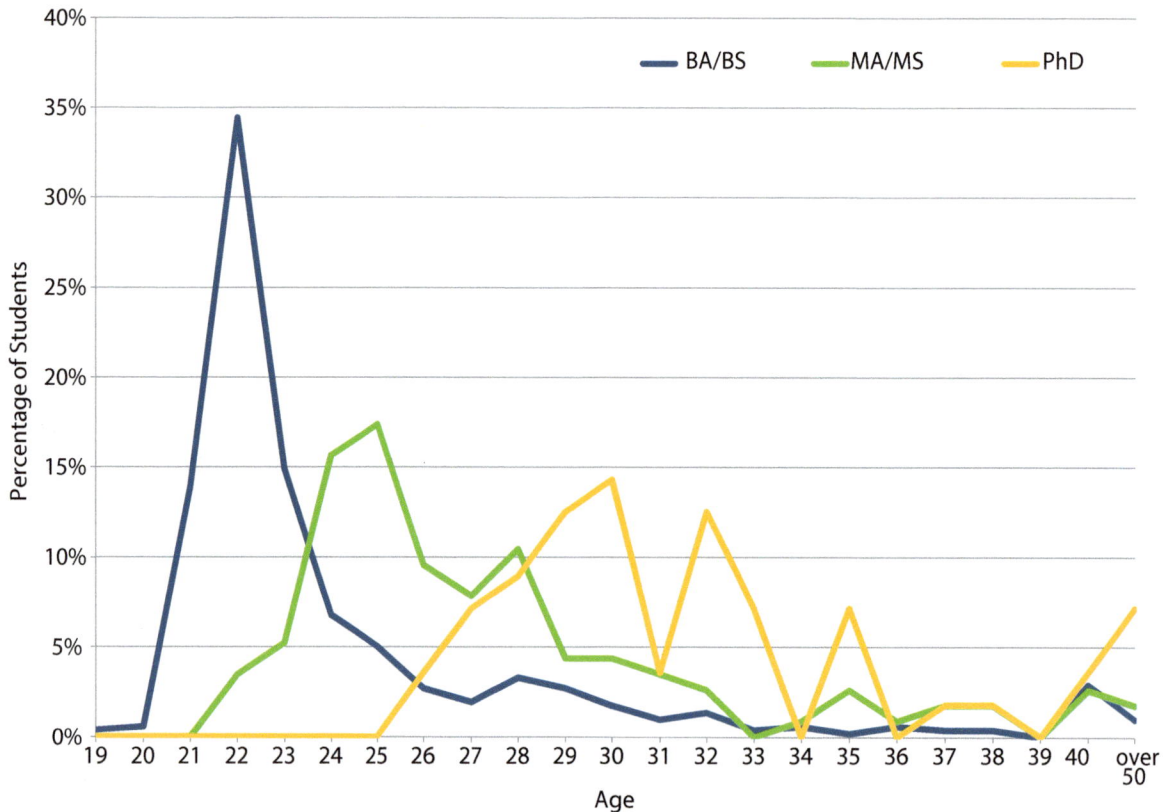

Citizenship of graduating students

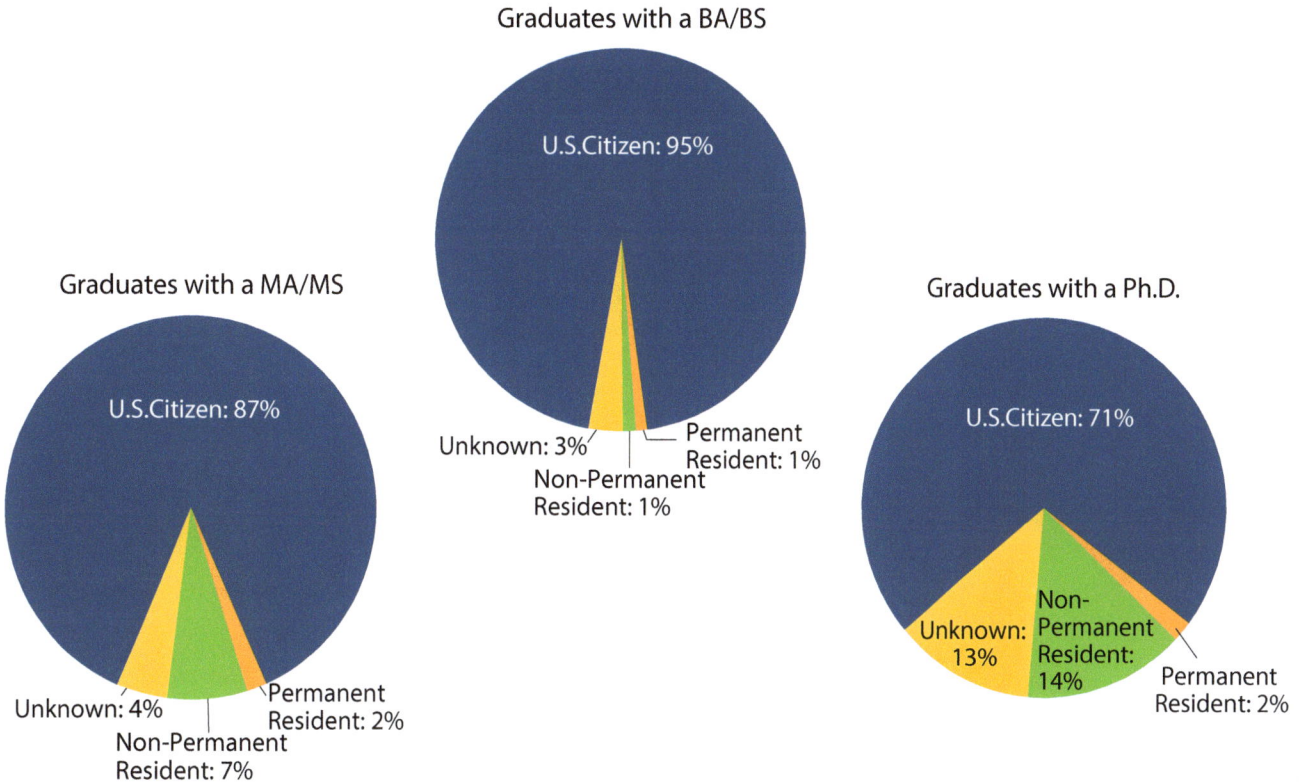

Graduates with a BA/BS

U.S.Citizen: 95%

Unknown: 3%
Non-Permanent Resident: 1%
Permanent Resident: 1%

Graduates with a MA/MS

U.S.Citizen: 87%

Unknown: 4%
Non-Permanent Resident: 7%
Permanent Resident: 2%

Graduates with a Ph.D.

U.S.Citizen: 71%

Unknown: 13%
Non-Permanent Resident: 14%
Permanent Resident: 2%

Race/ethnicity of graduating students

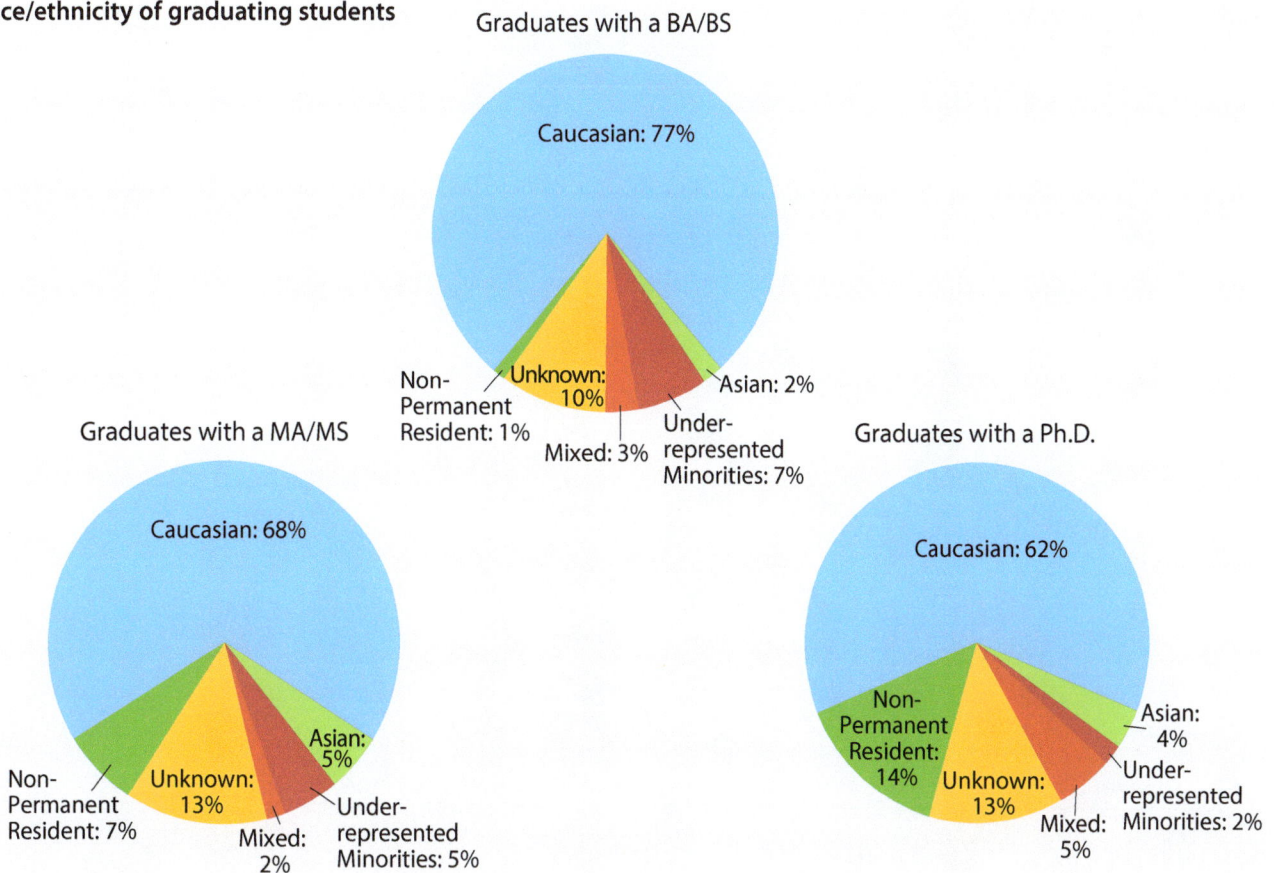

Graduates with a BA/BS

Caucasian: 77%

Non-Permanent Resident: 1%
Unknown: 10%
Mixed: 3%
Under-represented Minorities: 7%
Asian: 2%

Graduates with a MA/MS

Caucasian: 68%

Non-Permanent Resident: 7%
Unknown: 13%
Mixed: 2%
Under-represented Minorities: 5%
Asian: 5%

Graduates with a Ph.D.

Caucasian: 62%

Non-Permanent Resident: 14%
Unknown: 13%
Mixed: 5%
Under-represented Minorities: 2%
Asian: 4%

Quantitative Skills and Geoscience Background of the Graduating Students

This section examines graduates' educational background, such as quantitative rigor, the role of K-12 experiences, and the importance of two-year colleges.

The students were asked to select all the quantitative courses they have taken at a two-year or four-year institution. As in 2013, most bachelor's and master's graduates took Calculus II. A smaller percentage took higher level quantitative courses, and this percentage represents generally the same cohort of students taking those higher level classes. However, there is an overall drop in the percentages of graduates at all degree levels taking Statistics, as well as the other courses beyond Calculus II, in 2014. When looking at the classifications of the institutions these students attended, it is clear that most of the students that took quantitative courses beyond Calculus I attended schools classified as high and very high research institutions.

Along with identifying the quantitative courses taken, students were asked which additional core physical science courses they took during their postsecondary education. In 2014, the percentage of master's graduates that completed a chemistry course increased compared to 2013, and the doctoral graduates taking either a calculus or algebra-based physics course decreased in 2014 compared to 2013.

Students were asked if they took an earth science course in high school and if they attended a two-year college for at least a semester before receiving their degree. The percentages of geoscience graduates that took an earth science course in high school was just under 50 percent in 2014, While these percentages are a little lower that in 2013, this trend still seems to go against anecdotal claims of a lack of exposure to earth science in high schools. There was also an increase in the percentages of graduates at all three levels that attended a two-year college for at least a semester, particularly master's graduates. Increases in attendance at two-year colleges among geoscience graduates could continue due to increased efforts of collaboration between two-year and four-year institutions to create pathways for two-year college students to transition into geoscience majors at four-year universities.

Quantatitive skills and knowledge gained while working towards degree

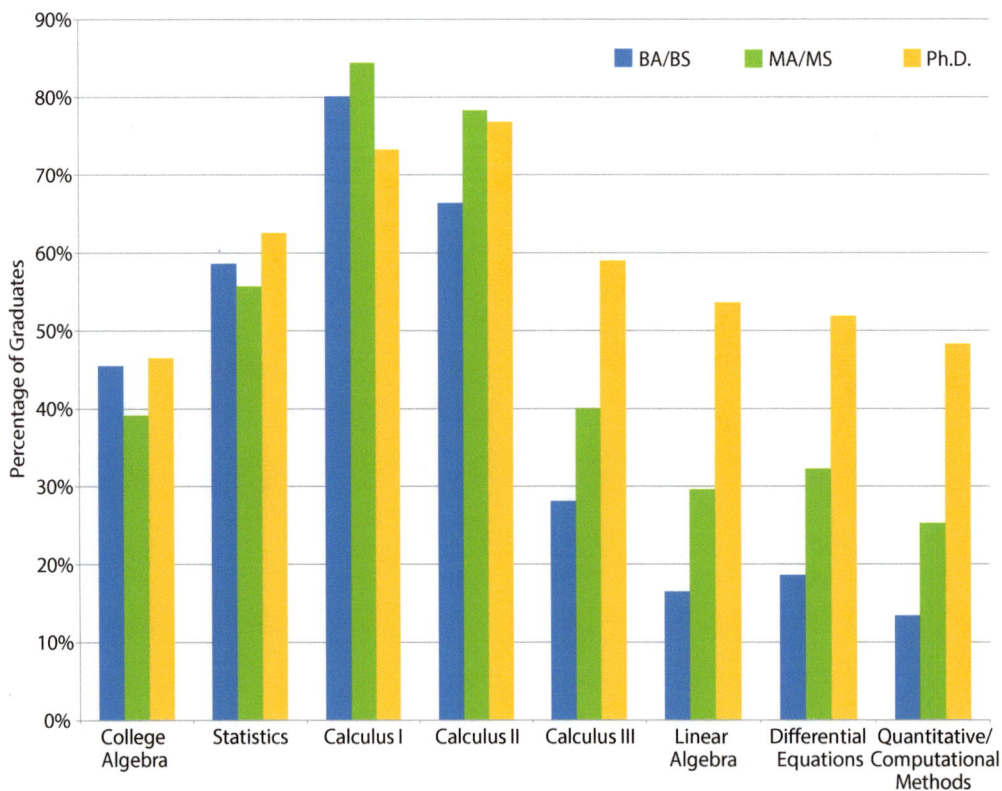

Quantitative skills and knowledge gained by graduates based on university classification**

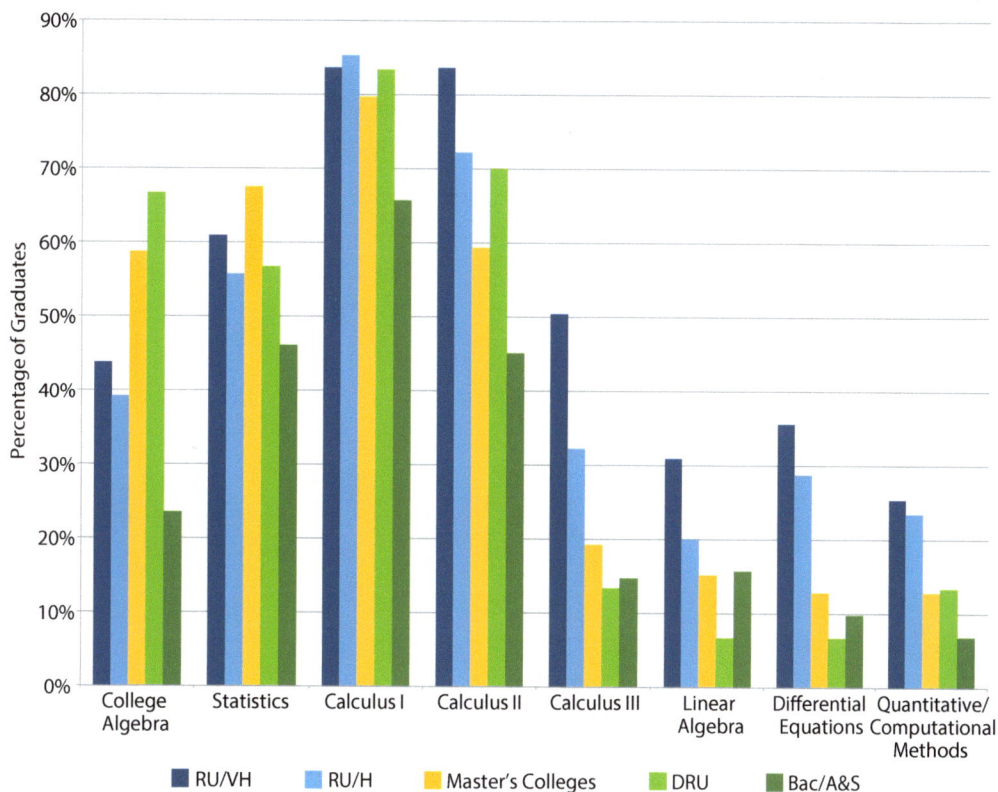

**See Appendix II for definitions of the Carnegie University Classification System

Quantitative skills and knowledge gained while working toward degree by gender

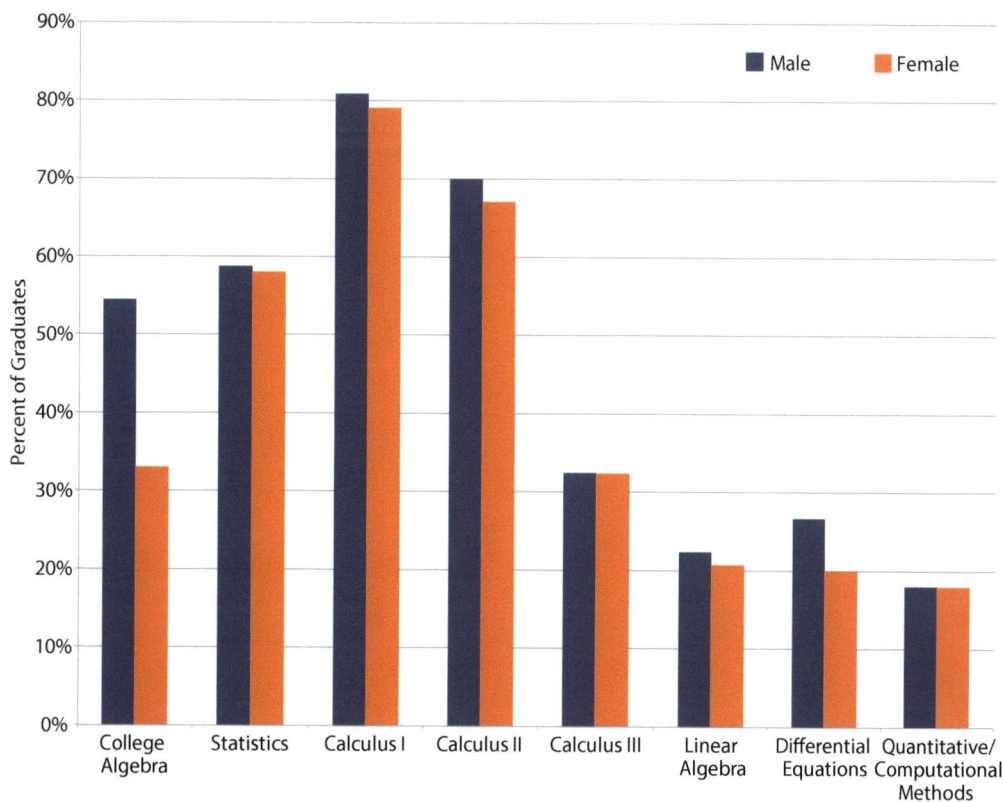

Percentage of graduates taking other core physical science courses

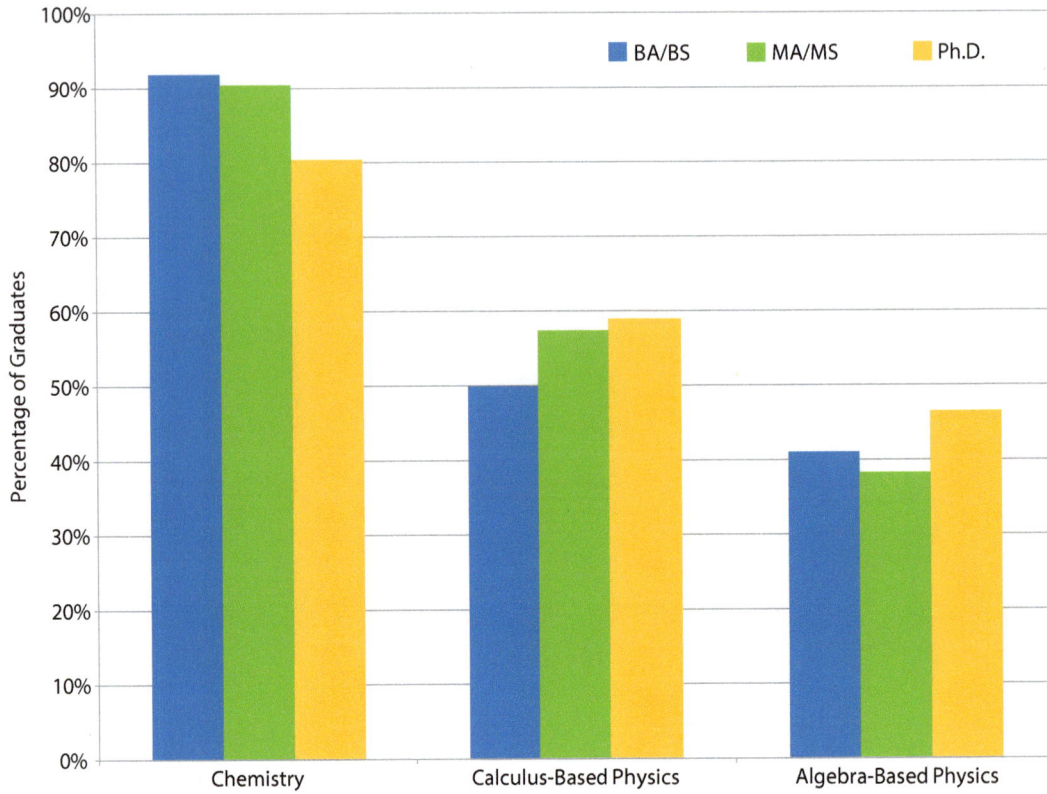

Photo by Rob Thomas from the AGI 2014 Life in the Field photo contest.

Graduates who took an earth science course in high school

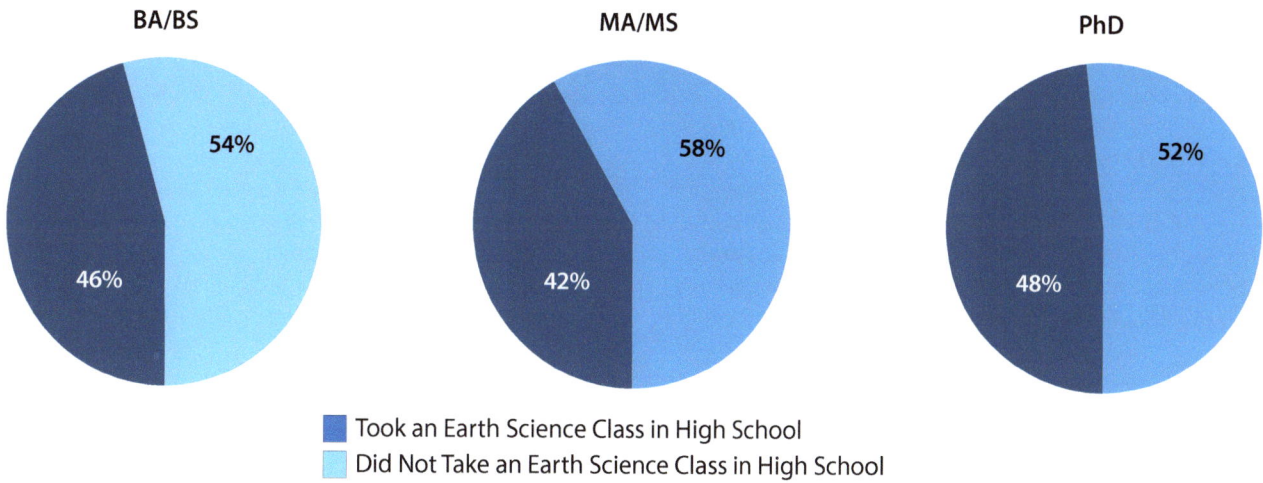

BA/BS

54%

46%

MA/MS

58%

42%

PhD

52%

48%

- ■ Took an Earth Science Class in High School
- ■ Did Not Take an Earth Science Class in High School

Graduates who attended a two-year college for at least 1 semester and took a geoscience course

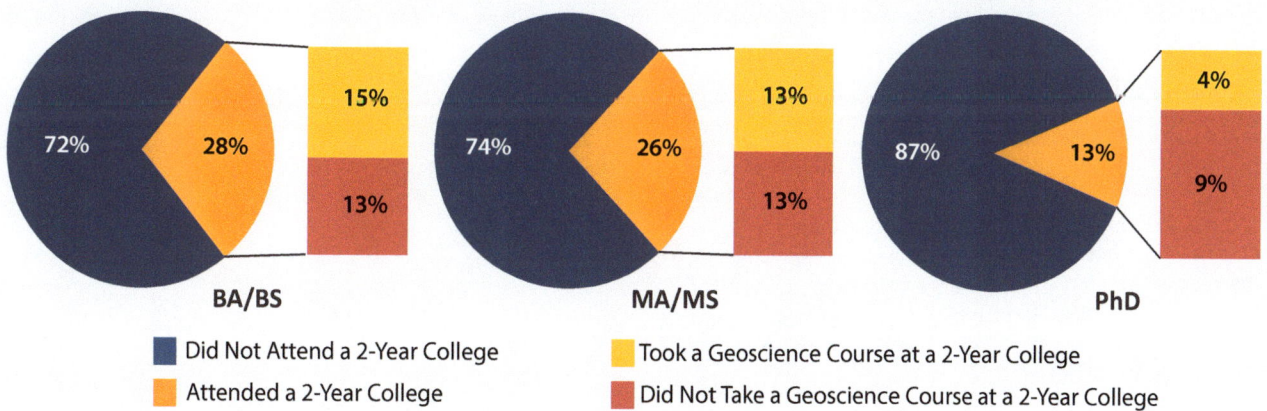

72% 28% 15% 13%

BA/BS

74% 26% 13% 13%

MA/MS

87% 13% 4% 9%

PhD

- ■ Did Not Attend a 2-Year College
- ■ Attended a 2-Year College
- ■ Took a Geoscience Course at a 2-Year College
- ■ Did Not Take a Geoscience Course at a 2-Year College

Choosing Geoscience as a Major

Graduates were asked which geoscience field they were getting a degree in, as well as the fields associated with any other postsecondary degrees. The chosen degree fields demonstrate the variety of disciplines related to the geosciences. Geology continues to be the most popular degree among undergraduates with students specializing in different fields more often in graduate school. The "Other Geosciences" categories represent degree fields such as economic geology, petrology, mineral physics, and paleomagnetism. For bachelor's graduates, the "Other" category also includes a few interdisciplinary degree fields specifically for a particular department that allows for an emphasis or minor in a geoscience field. For master's graduates, 63 percent switched their degree field within the geosciences for their graduate work and 19 percent had earned an undergraduate degree in a field outside of the geosciences. For doctoral graduates, 68 percent switched into a different geoscience degree field for their graduate work and 23 percent earned an undergraduate degree in a field outside of the geosciences.

In 2014, the majority of graduates at the bachelor's and master's levels chose to major in the geosciences at some point during their undergraduate educations. This trend was also seen in 2013, which highlights the importance of undergraduate geoscience courses to the recruitment of majors. However, in 2014, an equal percentage of doctoral graduates claimed to have chosen a geoscience major before beginning their postsecondary education and during their undergraduate education. While the majority of graduates at all levels indicated the intellectual engagement of geoscience as the reason for choosing their major, the students spoke of different intellectual draws to the field. Among bachelor's graduates, 20 percent found the college coursework and experiences engaging and 18 percent indicated an interest in the outdoors within the overall reason of intellectual engagement. Bachelor's graduates also cited the career opportunities and job security (24%) and the influence of a role model within the department, either a faculty member or fellow student, (21%) as reasons for majoring in the geosciences. Among master's graduates, 21 percent claimed an overall interest in science, 17 percent indicated an interest in the outdoors, and 16 percent found the college coursework and experiences interesting within the claim of intellectual engagement. Twenty-three percent of master's graduates also cited the career opportunities and job security as a reason to major in the geosciences. Among doctoral graduates, 18 percent were drawn to the interdisciplinary nature of the geosciences, 16 percent cited an overall interest in science, and 14 percent indicated both liking the outdoors and specifically the field work associated with the geosciences within the claim of intellectual engagement. Doctoral graduates also cited the importance of a role model within the department (14%) as a reason for majoring in the geosciences. Other reasons mentioned by graduates include the societal impact of the geosciences, the applicability of the geosciences to everyday life, and pre-college coursework.

When students decide to major in the geosciences

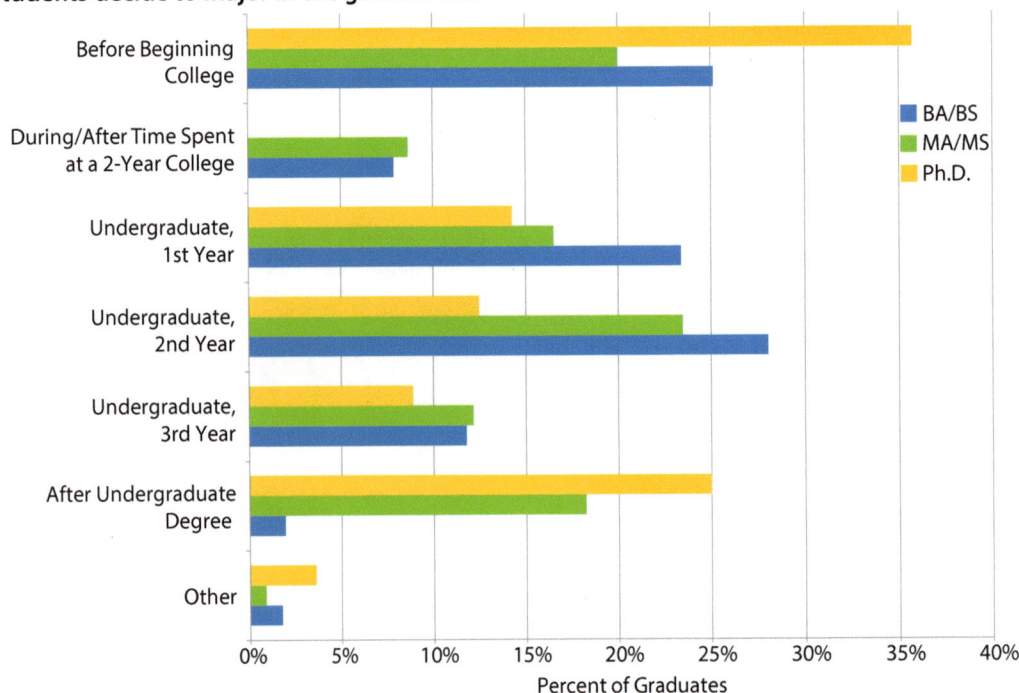

Chosen geoscience degree fields

Bachelor's Degree Graduates

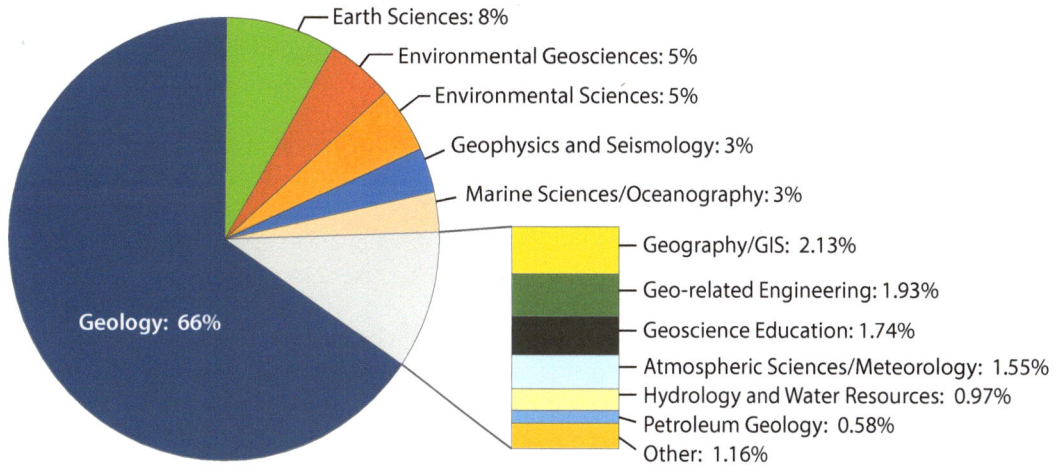

- Earth Sciences: 8%
- Environmental Geosciences: 5%
- Environmental Sciences: 5%
- Geophysics and Seismology: 3%
- Marine Sciences/Oceanography: 3%
- Geology: 66%
- Geography/GIS: 2.13%
- Geo-related Engineering: 1.93%
- Geoscience Education: 1.74%
- Atmospheric Sciences/Meteorology: 1.55%
- Hydrology and Water Resources: 0.97%
- Petroleum Geology: 0.58%
- Other: 1.16%

Master's Degree Graduates

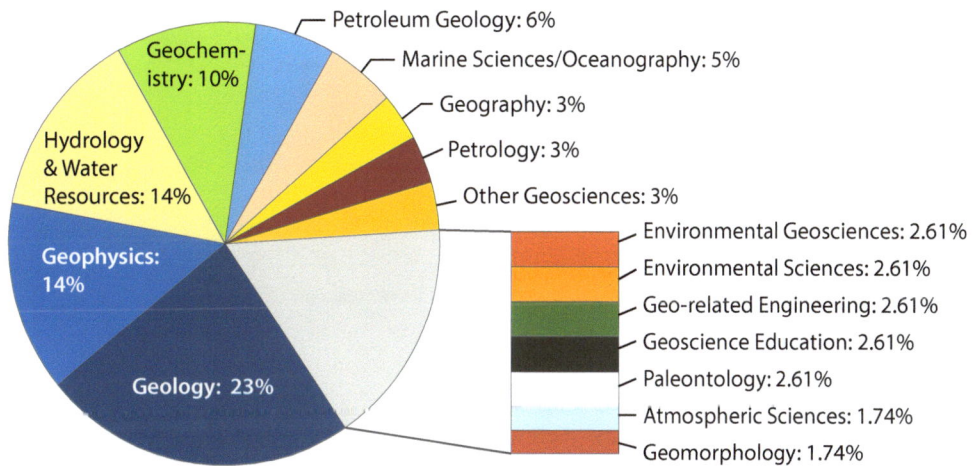

- Geochemistry: 10%
- Petroleum Geology: 6%
- Marine Sciences/Oceanography: 5%
- Geography: 3%
- Petrology: 3%
- Other Geosciences: 3%
- Hydrology & Water Resources: 14%
- Geophysics: 14%
- Geology: 23%
- Environmental Geosciences: 2.61%
- Environmental Sciences: 2.61%
- Geo-related Engineering: 2.61%
- Geoscience Education: 2.61%
- Paleontology: 2.61%
- Atmospheric Sciences: 1.74%
- Geomorphology: 1.74%

Doctoral Degree Graduates

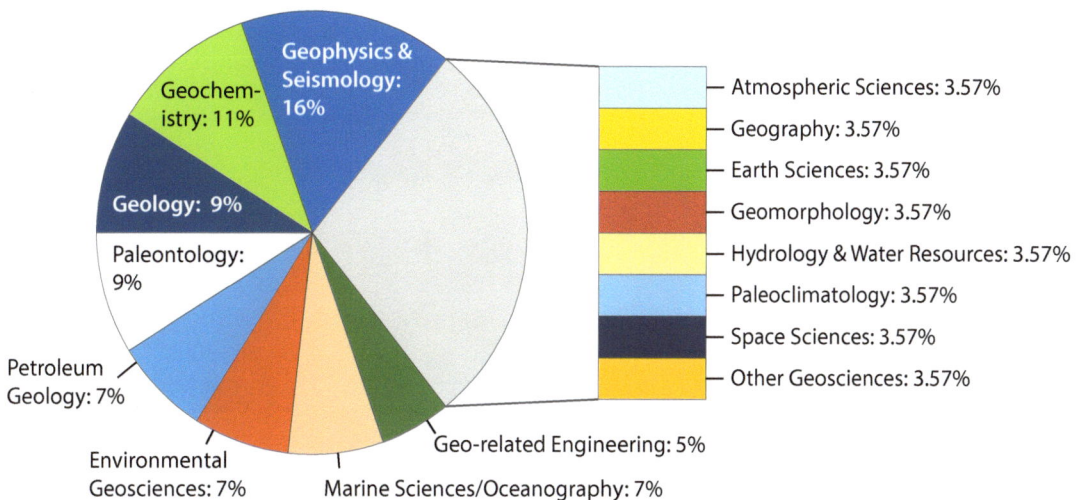

- Geophysics & Seismology: 16%
- Geochemistry: 11%
- Geology: 9%
- Paleontology: 9%
- Petroleum Geology: 7%
- Environmental Geosciences: 7%
- Marine Sciences/Oceanography: 7%
- Geo-related Engineering: 5%
- Atmospheric Sciences: 3.57%
- Geography: 3.57%
- Earth Sciences: 3.57%
- Geomorphology: 3.57%
- Hydrology & Water Resources: 3.57%
- Paleoclimatology: 3.57%
- Space Sciences: 3.57%
- Other Geosciences: 3.57%

Ancillary Factors Supporting the Degree

Graduates were asked about their experiences while working towards their degree. As in 2013, the majority of bachelor's graduates and doctoral graduates, 67 percent and 54 percent respectively, did not hold an internship during their postsecondary education. There was also an increase in the percentage of master's graduates that did not participate in an internship during their postsecondary education from 38 percent in 2013 to 54 percent in 2014. This continued trend is worrisome due to invaluable experiences and networking that occur during internship experiences. Also, among the graduates that participated in an internship, 74 percent of bachelor's graduates, 81 percent of master's graduates, and 73 percent of doctoral graduates rated their internships as "very important" for their academic and professional development. This is reinforced by the graduates that found a job within the geosciences at time of graduation. Nineteen percent of bachelor's graduates, 51 percent of master's graduates, and 67 percent of doctoral graduates that completed an internship during their postsecondary education secured a job in the geosciences at the time of graduation.

When asked about the types of financial aid used to fund their education, over 80 percent of graduates at all levels had some sort of financial aid while working towards their degree. In 2014, there was a decrease in the percentage of graduates at the bachelor's and doctoral levels that used student loans to fund their education, but student loans are still utilized by graduates at all levels. Also, there was a decrease in the percentage of master's and doctoral graduates that held a research assistantship compared to 2013, and an increase in the percentage of doctoral graduates that held a teaching assistantship compared to 2013.

Graduates were also asked about their involvement with geoscience membership organizations. AGI is a federation of 49 different geoscience membership societies, including societies such as the American Geophysical Union, the American Institute of Professional Geologists, and the Geological Society of America. Professional societies can be useful tools for success as an early-career geoscientist, but the percentages of participation in these organizations, particularly among students finishing their graduate degrees, were surprisingly low. Only 21 percent of bachelor's graduates, 39 percent of master's graduates, and 39 percent of doctoral graduates mentioned association with one of these organizations.

Number of internships held by graduating students

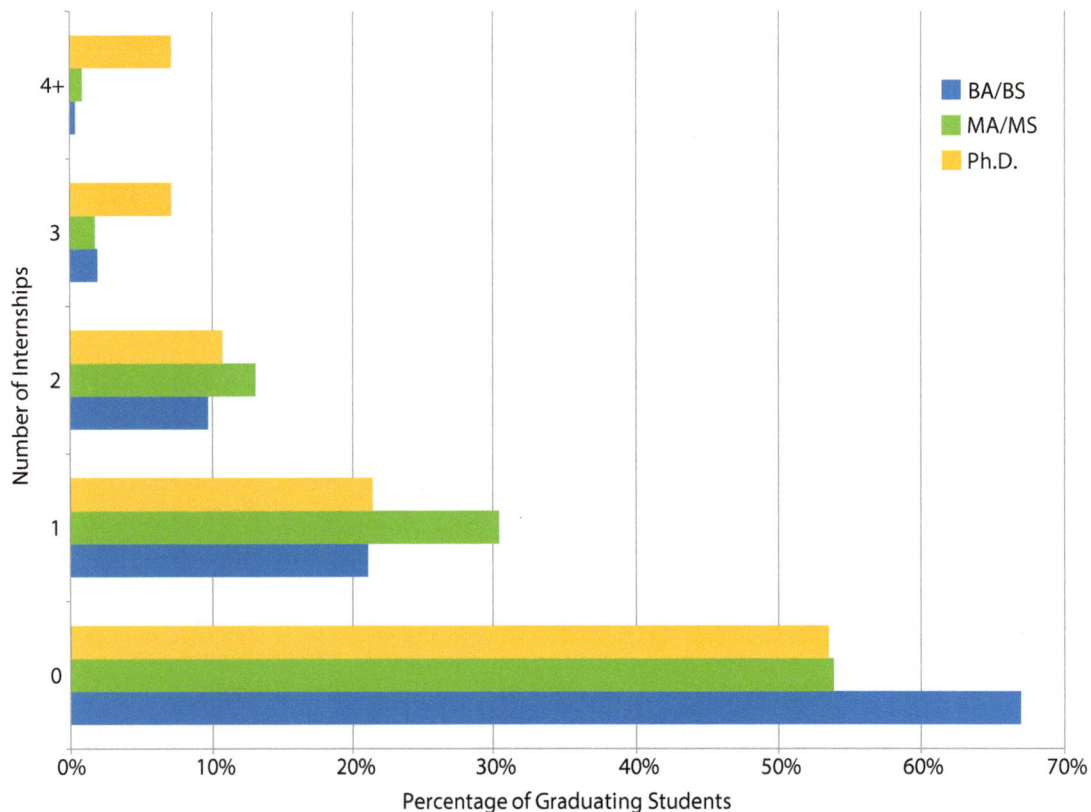

Types of financial aid used by graduating students while working towards a degree

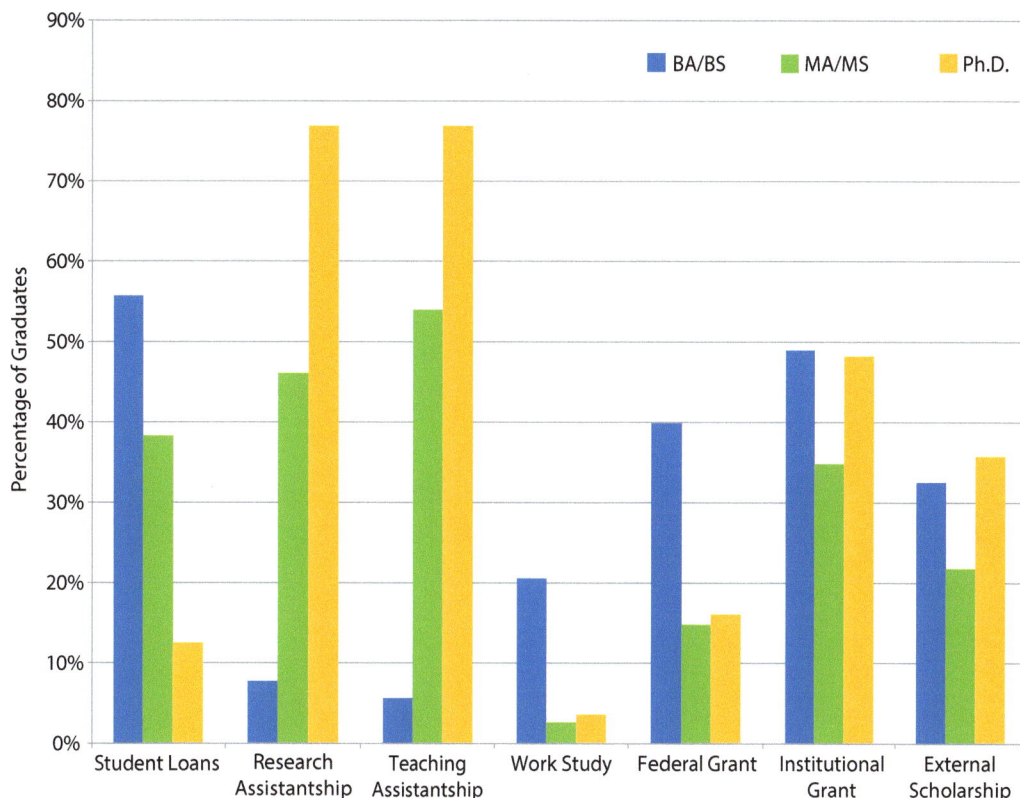

Participation in geoscience clubs

	BA/BS	MA/MS	Ph.D.
Associated with a geoscience-related club/organization	65%	76%	70%
Participated in department-level geoscience club	50%	53%	34%
Member of an AGI Member Society	21%	39%	39%
Member of an Honor Society	12%	10%	9%

Average GPA

	BA/BS	MA/MS	Ph.D.
Average years to degree completion	3.19	2.45	4.84
Average overall GPA	3.77	3.75	3.89
Average geoscience GPA	4.00	3.78	3.91

Field Experiences

Clear definitions were set to distinguish between field camp, field courses, and field experiences. A field camp was defined as an academic program lasting four or more weeks that is primarily focused on field tools and methods. A field course was defined as a course with a field component primarily covering field methods and experimentation that utilized at least half of the total class time. A field experience was defined as any course that contained a field component, such as a field trip, field work, or other time in the field, that is not included in the definitions for field camp or field course.

Nearly every graduate that took the survey had at least one field experience while working towards their degree. In 2014, there was a slight increase in the percentage of bachelor's graduates that participated in field camp and a slight decrease in the percentage of bachelor's graduates that still have the desire to participate in field camp compared to 2013. However, in 2014, we see the opposite trends among master's graduates. There is also a gender difference in field camp participation with approximately 10 percent more men attending field camp before graduation than women. When asked about the importance of these field experiences to the graduates' academic and professional development, field camps, field courses, and field experiences were all rated as "very important" by the majority of graduates. However, this rating was the highest for all degree levels for field experiences, with 86 percent of bachelors' graduates, 83 percent of master's graduates, and 80 percent of doctoral graduates finding these experiences "very important" to their academic and professional development.

Graduates' participation in field experiences was also broken down by the Carnegie Classification of the institution (see Appendix II). While field experiences are prevalent at all the participating universities, research universities continue to provide more expansive access to field camp experiences.

Student participation in field experiences based on university classification**

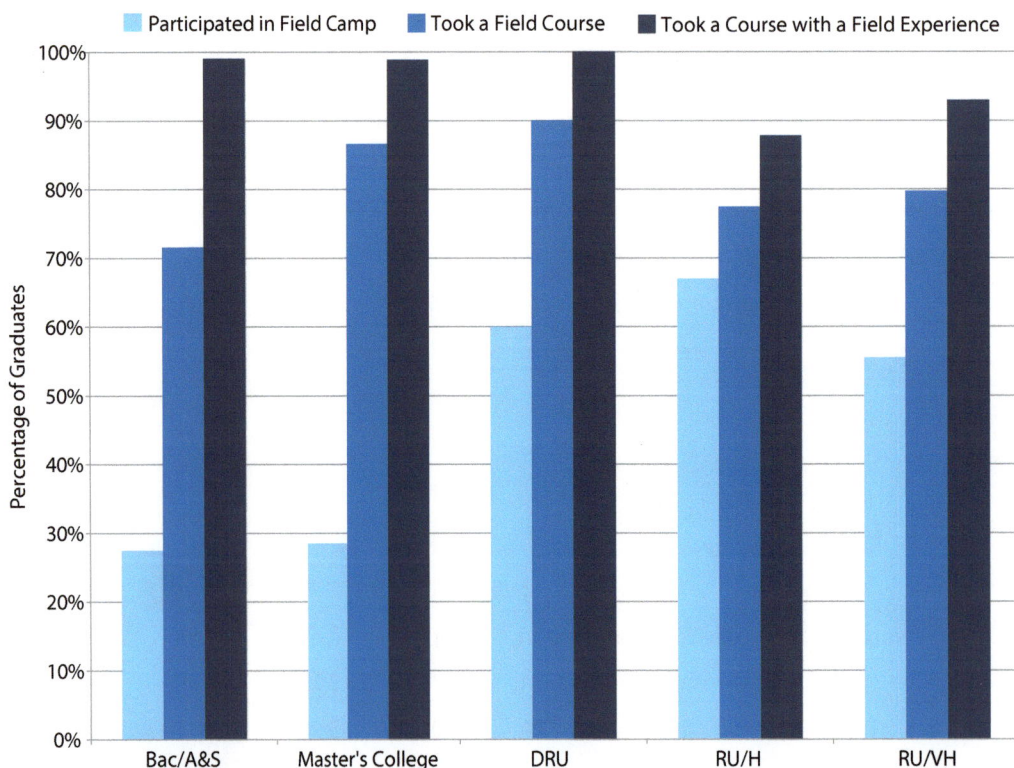

**See Appendix II for definitions of the Carnegie University Classification System

Graduating students who have participated in field camp

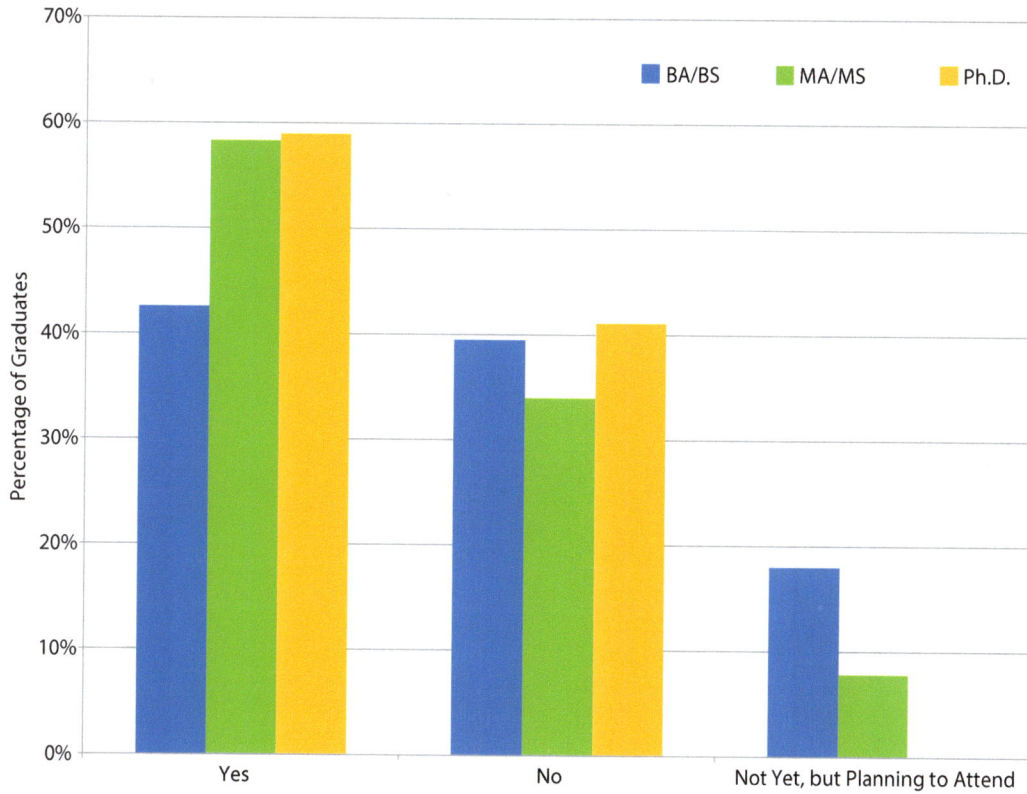

Graduating students who have participated in field camp by gender

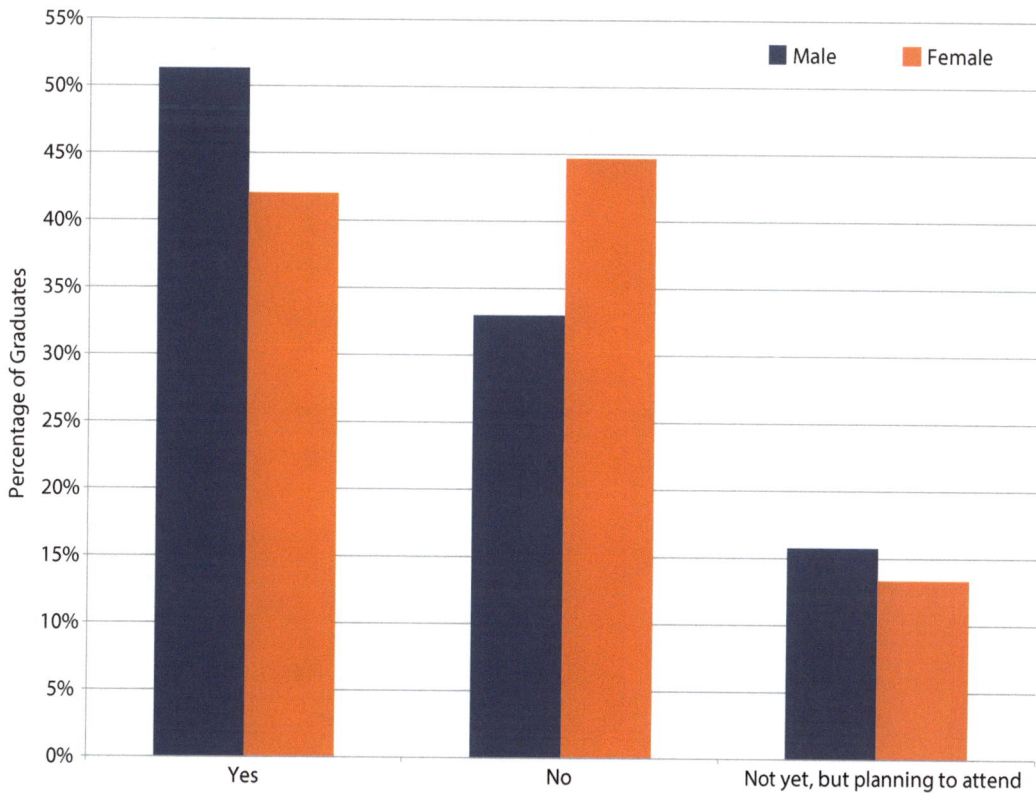

Graduates who have taken one or more field courses

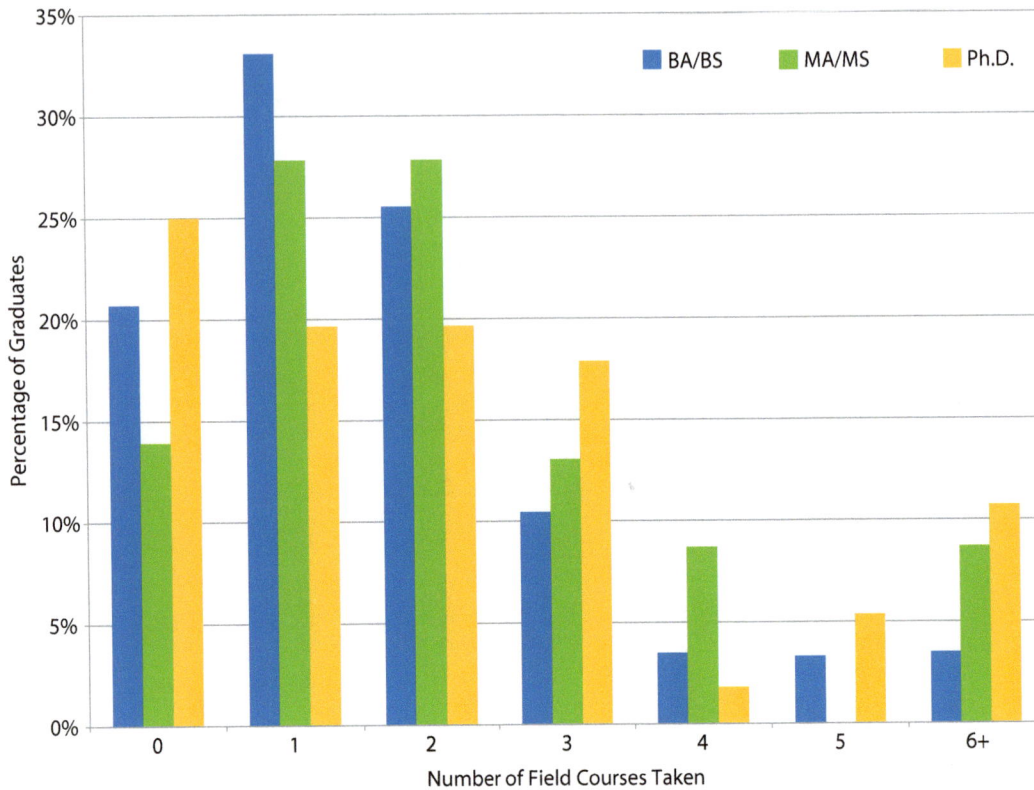

Graduates who have taken one or more courses with a field experience

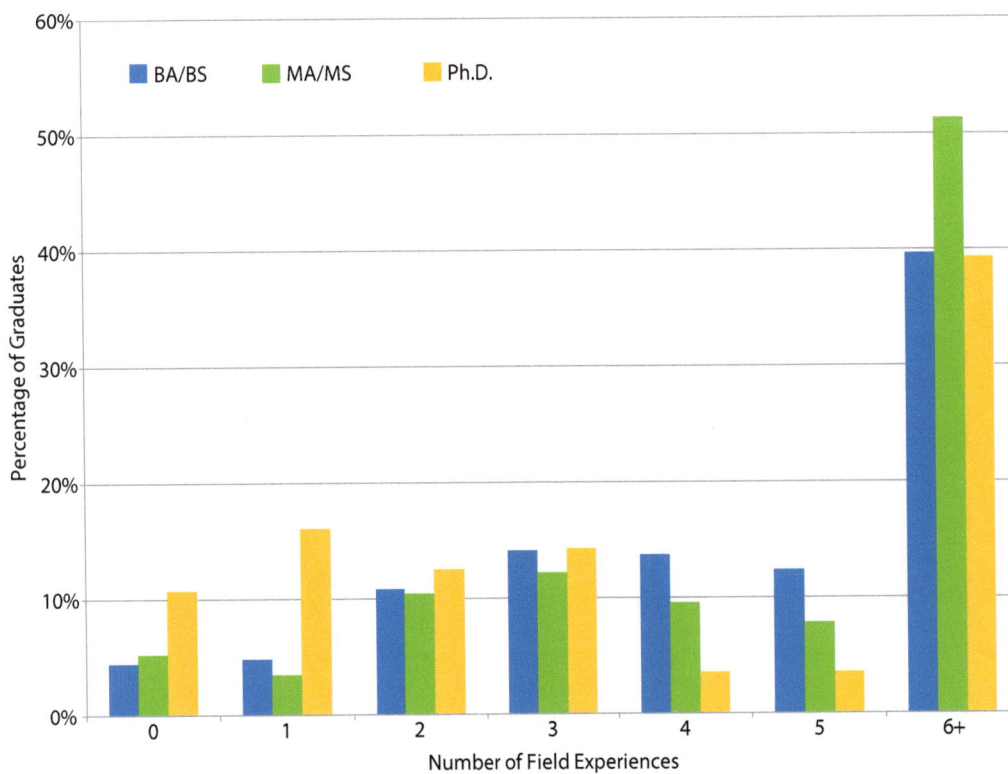

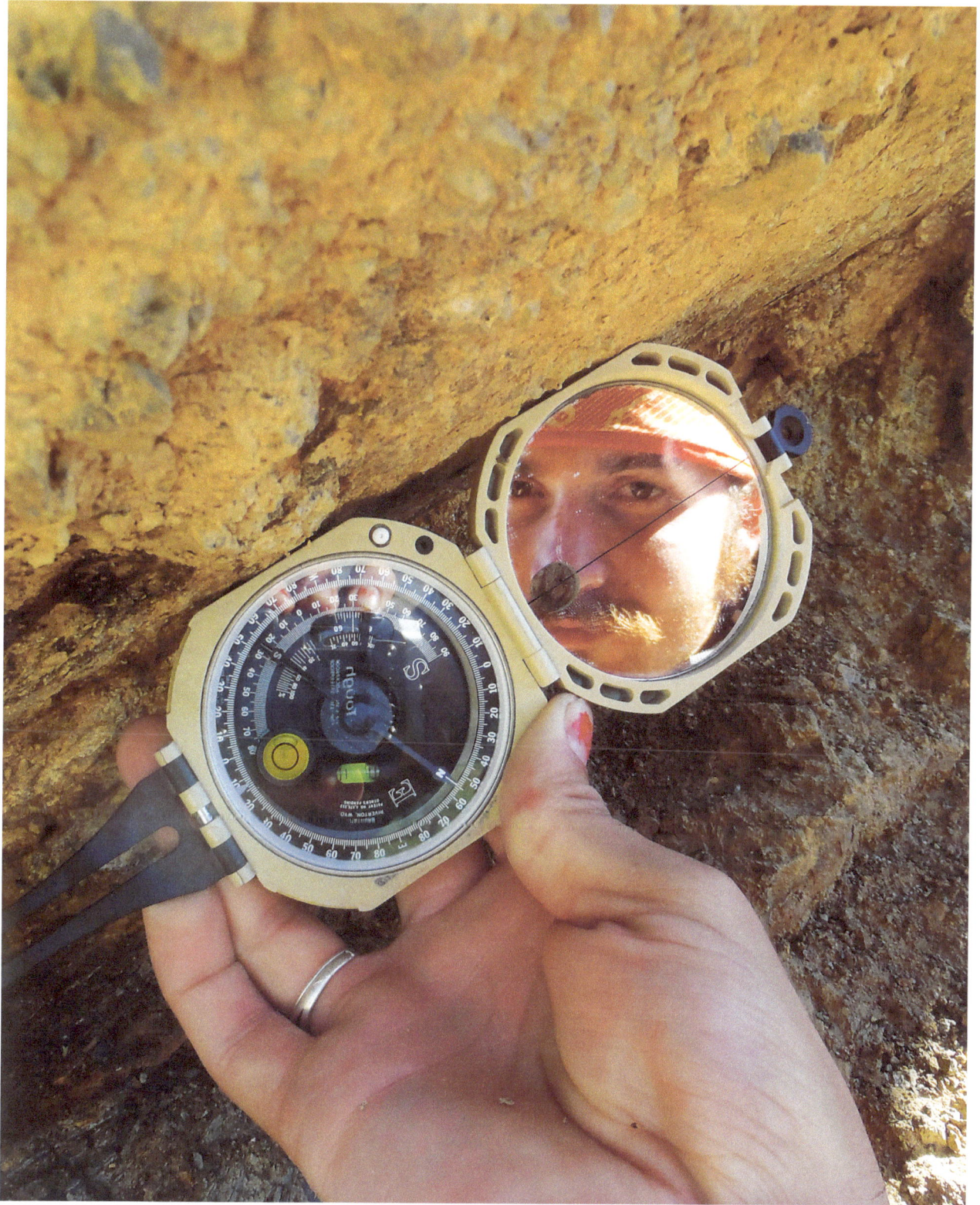

Photo by Samuel Castonguay from the AGI 2014 Life in the Field photo contest.

Research Experiences

The graduates were asked about their research experiences while working towards their degrees. If they indicated participation in at least one research experience, the graduates were then asked about their participation in faculty-directed research and self-directed research. If they indicated participation in self-directed research, they were then asked to identify the basic research methodology used to conduct their research.

In 2014, there was a slight increase in the percentage of bachelor's graduates participating in at least one research activity while working towards their degree compared to 2013. Additionally, there was an increase at all degree levels in the percentage of graduates participating in self-directed research in 2014 compared to 2013.

In 2014, undergraduate students continued to use field-based and lab-based research methods more often, but 55 percent of bachelor's graduates also used literature-based methods, which is an 11 percent increase from 2013. However, among graduate students finishing their degree in 2014, there was a shift compared to 2013 with an increased percentage of students utilizing field-based methods and a decreased percentage of students using lab-based methods. The specific methods used have dependency on the field of study.

When asked about the importance of research experiences to the graduates' academic and professional development, research experiences were rated "very important" by 83 percent of bachelor's graduates, 88 percent of master's graduates, and 96 percent of doctoral graduates.

Research methods utilized by graduates in their self-directed research

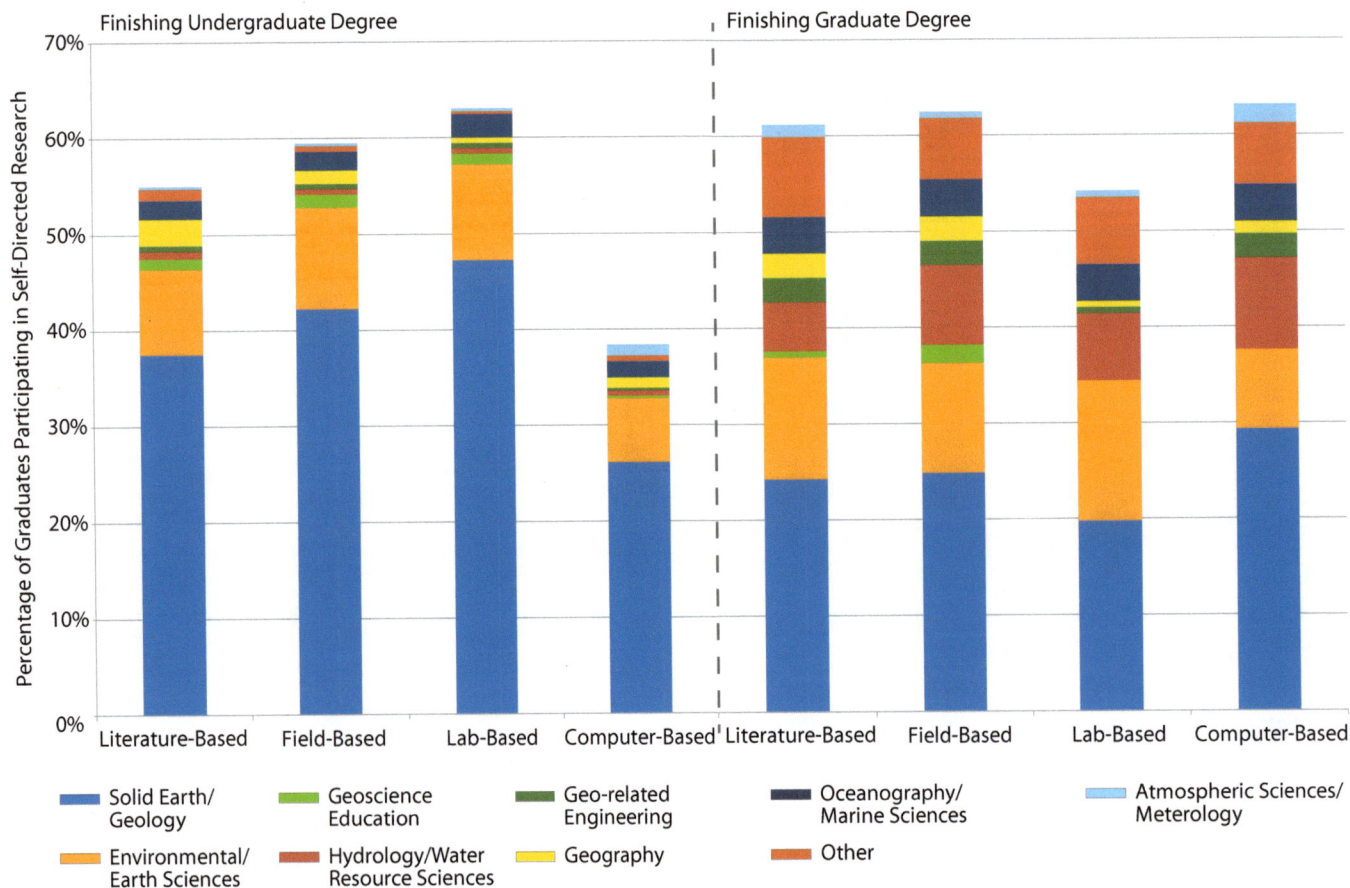

Legend:
- Solid Earth/Geology
- Geoscience Education
- Geo-related Engineering
- Oceanography/Marine Sciences
- Atmospheric Sciences/Meterology
- Environmental/Earth Sciences
- Hydrology/Water Resource Sciences
- Geography
- Other

Graduates with one or more research experiences

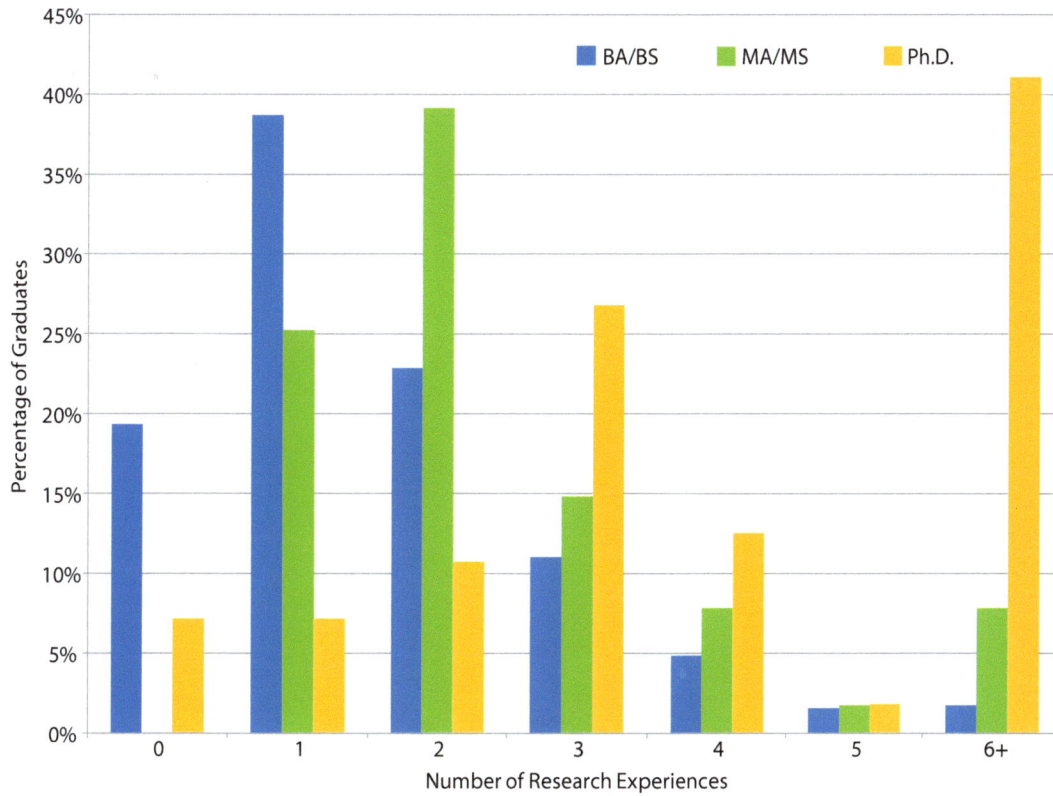

Student participation in faculty-directed and self-directed research

	BA/BS	MA/MS	Ph.D.
Faculty-Directed Research	56%	70%	91%
Self-Directed Research	70%	90%	96%

Research methods utilized by graduates in their self-directed research by gender

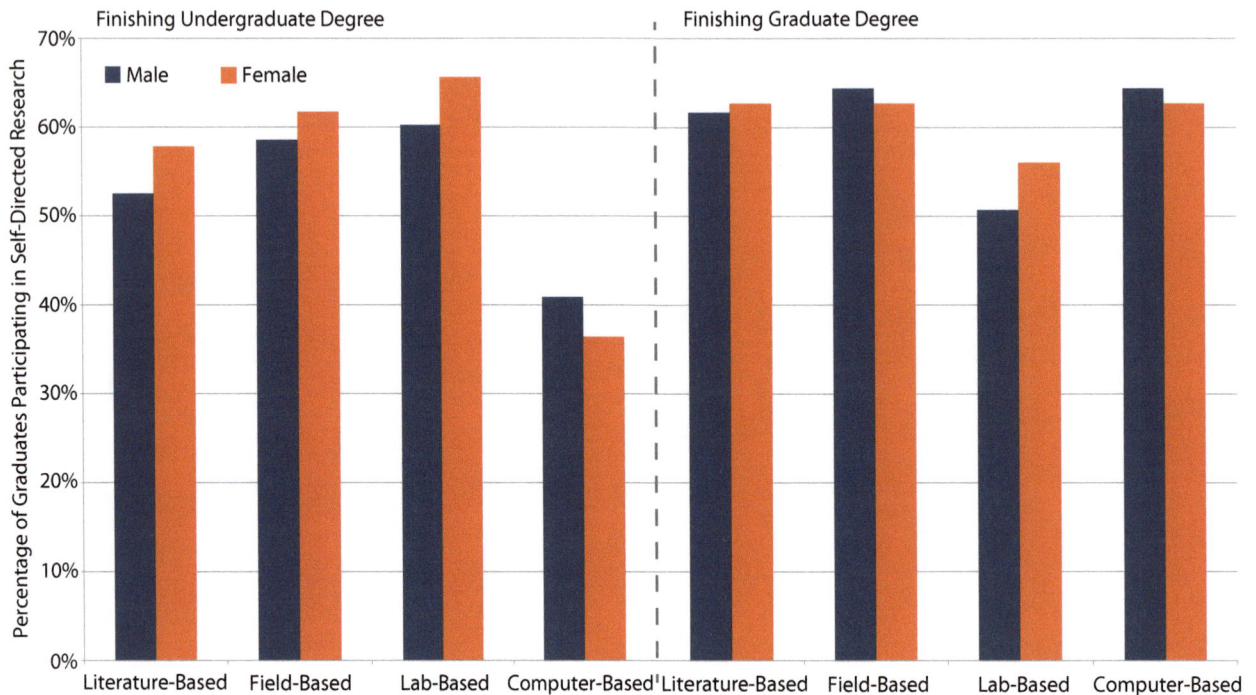

Finishing Undergraduate Degree / **Finishing Graduate Degree**

Legend: Male, Female

Student participation in research based on university classification**

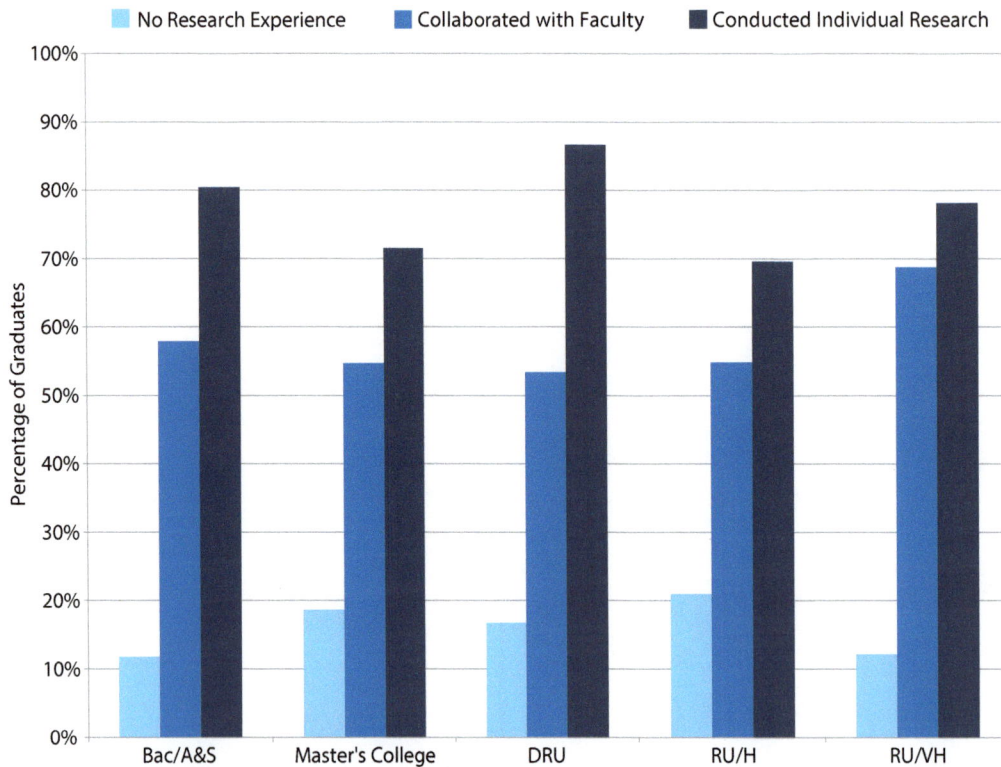

Legend: No Research Experience, Collaborated with Faculty, Conducted Individual Research

Categories: Bac/A&S, Master's College, DRU, RU/H, RU/VH

**See Appendix II for definitions of the Carnegie University Classification System

Photo by Ijeamaka Okechukwu from the AGI 2014 Life in the Field photo contest.

Future Plans: Working Toward a Graduate Degree

The graduates were asked if they have immediate plans to continue their education. Those indicating plans for a graduate degree after graduation were then asked to share the degree they would pursue and the degree field of interest for the degree.

In 2013, 38 percent of bachelor's graduates were planning to attend graduate school, which was higher than prior surveys that showed intent to continue education among new bachelor's graduates between 28 to 35 percent. In 2014, this percentage increased again to 42% of bachelor's graduates planning to attend graduate school in the immediate future. However, there is concern that geoscience graduate departments are reaching capacity, which will make it difficult for this increasing number of bachelor's graduates trying to continue their education. There was also an increase in the percentage of master's graduates planning to continue their education from 17 percent in 2013 to 26 percent in 2014.

New bachelor's graduates showed a varied array of intended graduate degree fields. The fields that fell into "Other Geosciences" included gemology, soil science, petrology, energy and earth resources, volcanology, mineral and energy economics, and environmental public health. The fields that fell into "Other" included public administration, museum studies, graphic design, disaster medicine and management, exercise and sport science, and zoology.

Unlike in 2013, the majority of graduate students wanting to obtain a second graduate degree intended to work towards another master's degree.

Students planning to attend graduate school after graduation

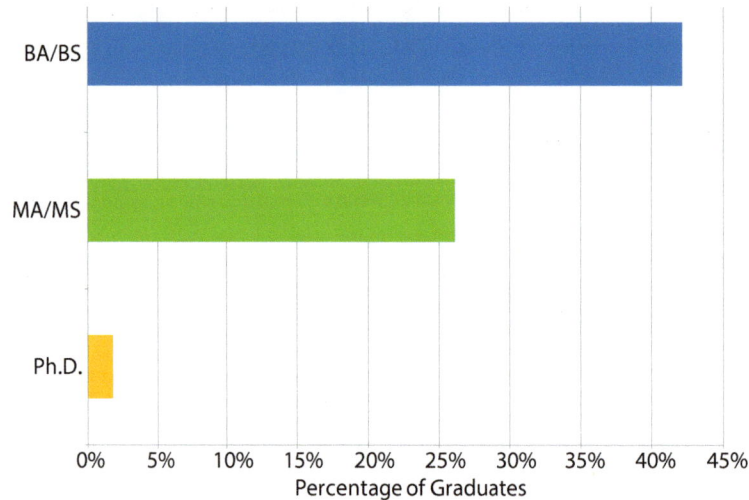

Students planning to attend graduate school after graduation by gender

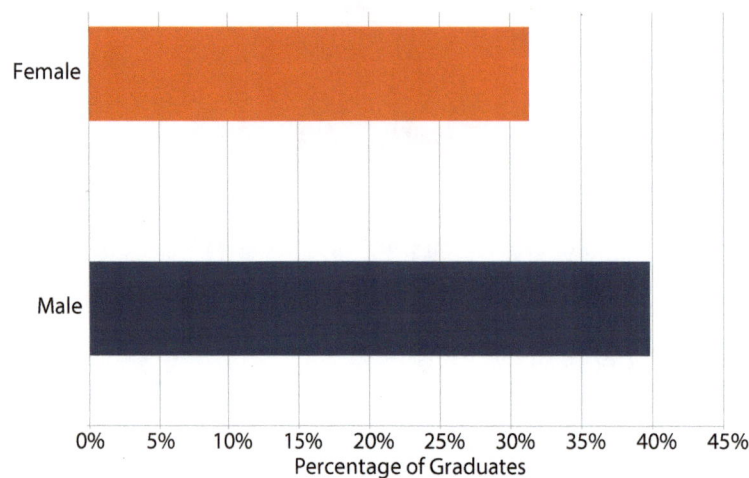

Students graduating with an undergraduate degree

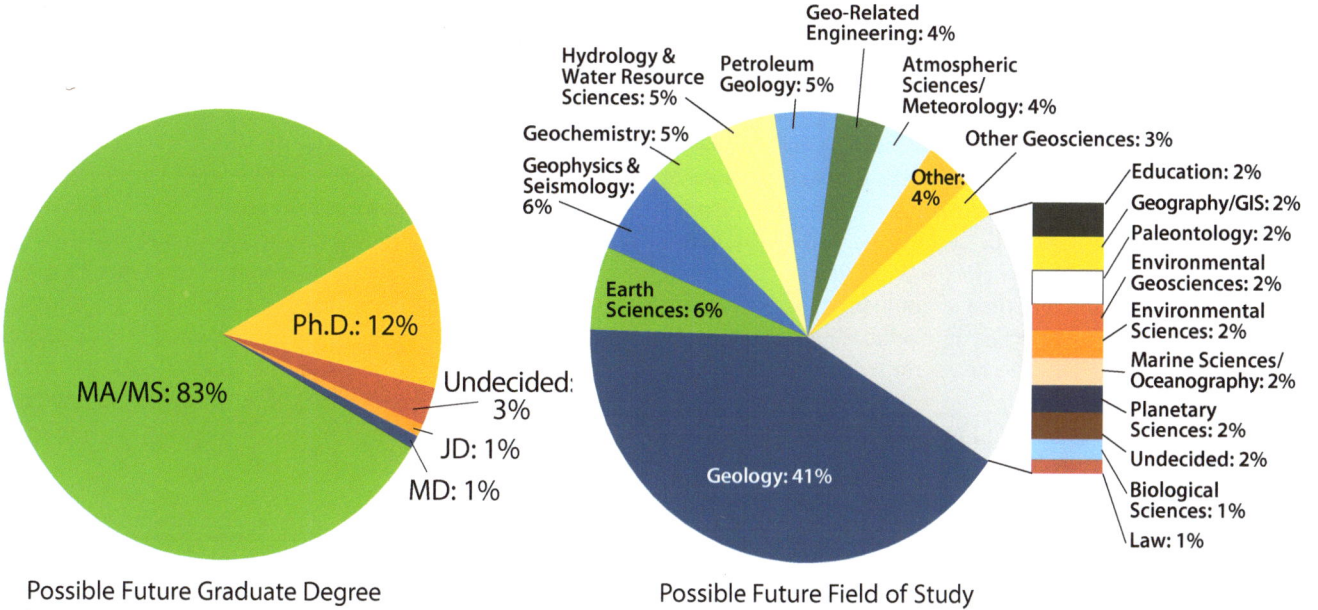

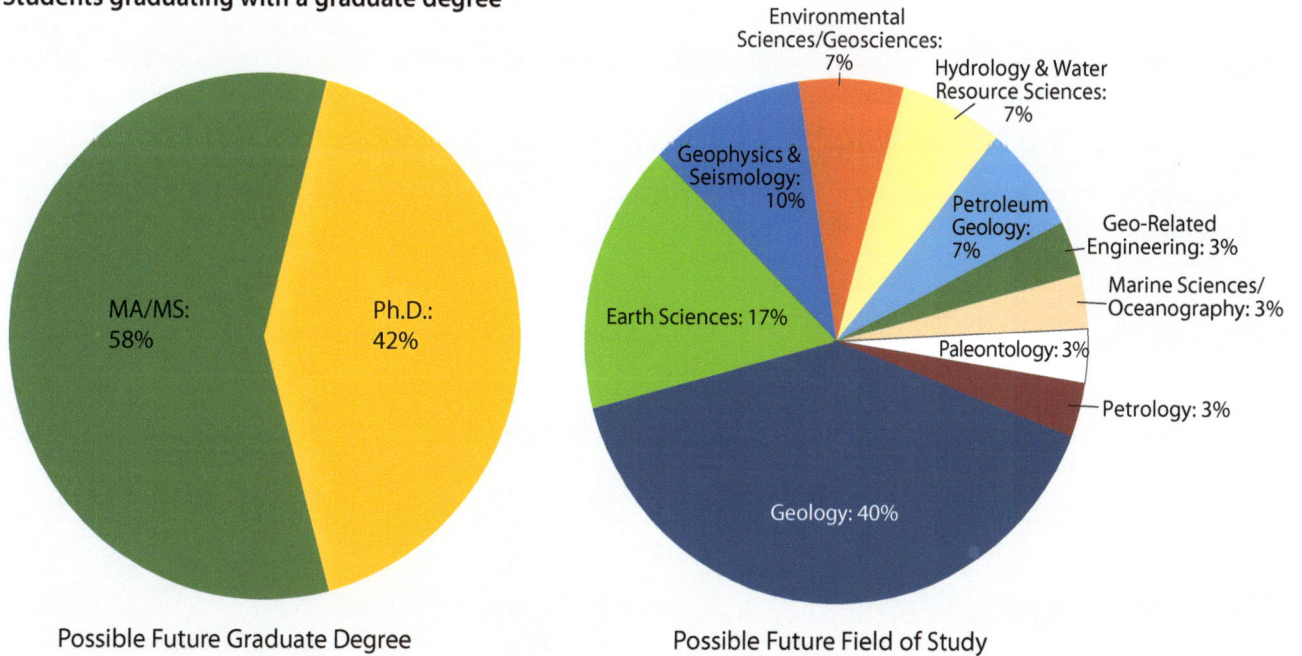

MA/MS: 83%
Ph.D.: 12%
Undecided: 3%
JD: 1%
MD: 1%

Possible Future Graduate Degree

Geo-Related Engineering: 4%
Hydrology & Water Resource Sciences: 5%
Petroleum Geology: 5%
Atmospheric Sciences/Meteorology: 4%
Geochemistry: 5%
Other Geosciences: 3%
Geophysics & Seismology: 6%
Other: 4%
Earth Sciences: 6%
Education: 2%
Geography/GIS: 2%
Paleontology: 2%
Environmental Geosciences: 2%
Environmental Sciences: 2%
Marine Sciences/Oceanography: 2%
Planetary Sciences: 2%
Undecided: 2%
Biological Sciences: 1%
Geology: 41%
Law: 1%

Possible Future Field of Study

Students graduating with a graduate degree

MA/MS: 58%
Ph.D.: 42%

Possible Future Graduate Degree

Environmental Sciences/Geosciences: 7%
Hydrology & Water Resource Sciences: 7%
Geophysics & Seismology: 10%
Petroleum Geology: 7%
Geo-Related Engineering: 3%
Earth Sciences: 17%
Marine Sciences/Oceanography: 3%
Paleontology: 3%
Petrology: 3%
Geology: 40%

Possible Future Field of Study

Future Plans: Working in the Geosciences

The graduates were asked if they had accepted or were seeking a job position within the geoscience workforce. If they had accepted a job, they were asked about these accepted job positions. Because the graduates take this survey right around graduation, it is not surprising that there are still relatively high percentages of graduates at all degree levels still seeking employment. In 2014, there was an increase in the doctoral graduates that had found a job position at time of graduation from 43 percent in 2013 to 70 percent. However, in 2014 there was a decrease in the bachelor's and master's students that found a job position at time of graduation, from 15 percent in 2013 to 12 percent for bachelor's graduates and 43 percent in 2013 to 35 percent for master's graduates.

Overall, the oil and gas industry continues to hire more graduates than any other industry, largely driven by effective on campus recruitment and highly competitive salaries. The top three industries hiring bachelor's graduates are the oil and gas industry, environmental services companies, and four-year universities. The top three industries hiring master's degrees are the oil and gas industries, environmental services companies, and state or local governments. The top three industries hiring doctoral graduates are four-year universities, the oil and gas industry, and research institutions. These jobs also tended to be primarily located in Texas.

As expected, most bachelor's graduates continued to find jobs with an annual salary between $20,000 and $60,000. The salary range for master's and doctoral students vary widely depending on the position. However, as in 2013, it appears the master's graduates tend to find jobs with higher annuals salaries than doctoral graduates. Also, as in 2013, every graduate making an annual salary of more than $90,000 found their job in the oil and gas industry. However, not every graduate with a job in the oil and gas industry has an annual salary above $90,000. Sixty-three percent of all graduates with a job in the geosciences mentioned receiving additional compensation from their employer as a bonus, moving expenses, etc. In 2014, this compensation was typically less than $10,000, with some graduates receiving more than $25,000.

Graduates that found geoscience employment were asked to identify the resources they used to find their job. Bachelor's graduates and doctoral graduates continued to rely largely on their personal contacts and faculty referrals, whereas master's graduates in 2014 relied on campus recruiting events and job fairs, personal contacts, and internet job boards or searches. Overall in 2014, these graduates found increased success using of the internet and less success through faculty referrals and campus recruiting events compared to 2013.

In 2014, the graduates at all degree levels still seeking employment in the geosciences showed interest in the same industries that tended to hire their peers.

The circular figure displays the connection between the degree fields of recent geoscience graduates in 2013 and 2014 (in color) to the industries where these geoscientists found their first job after graduation (in gray). The size of the bars along the outer edge of the circle represents the number of recent graduates that pursued a particular degree field and entered a particular industry. Each colored, inner ribbon connects a particular degree field with the various industries where graduates found jobs. The thickness of each ribbon is determined by the number of graduates within each degree field with a job in a particular industry. This visualization shows the variety of industries available to graduates with a geoscience degree, as well as the complexity of the workforce and knowledge needed in the distinct industries.

Photo by Judah Epstein from the AGI 2014 Life in the Field photo contest.

Graduating students seeking or have accepted a position within the geosciences

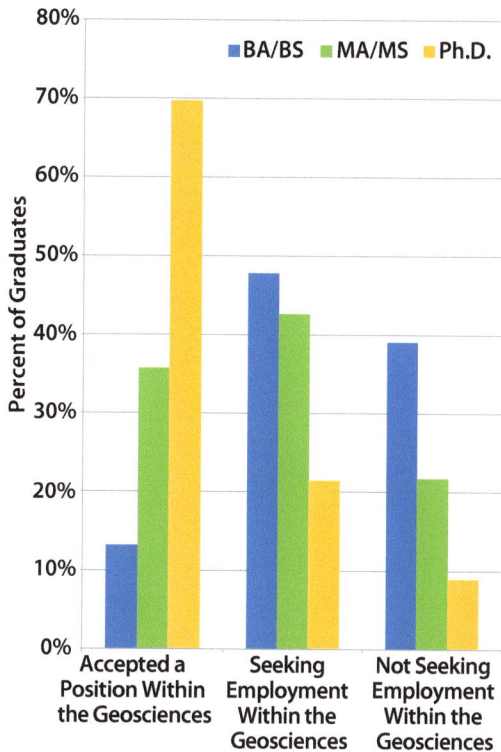

Graduating students seeking or have accepted a job within the geosciences by gender

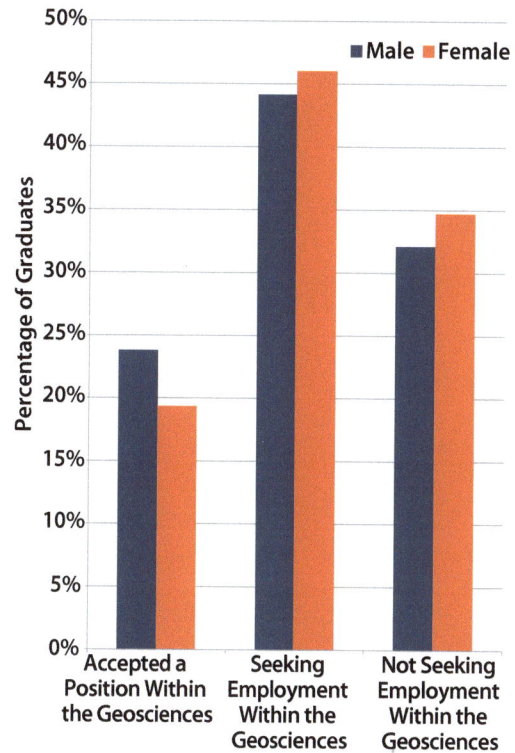

Industries where graduating students have accepted a job within the geosciences

Bachelor's Graduates

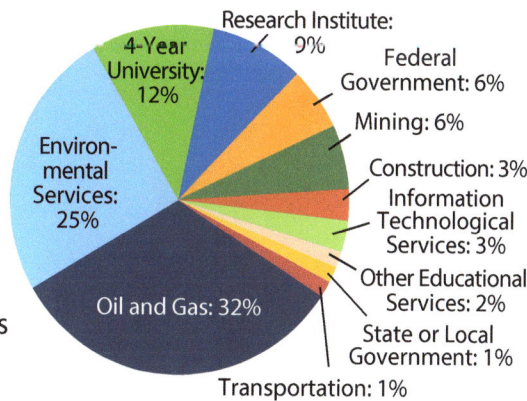

- Research Institute: 9%
- Federal Government: 6%
- Mining: 6%
- Construction: 3%
- Information Technological Services: 3%
- Other Educational Services: 2%
- State or Local Government: 1%
- Transportation: 1%
- Oil and Gas: 32%
- Environmental Services: 25%
- 4-Year University: 12%

Master's Graduates

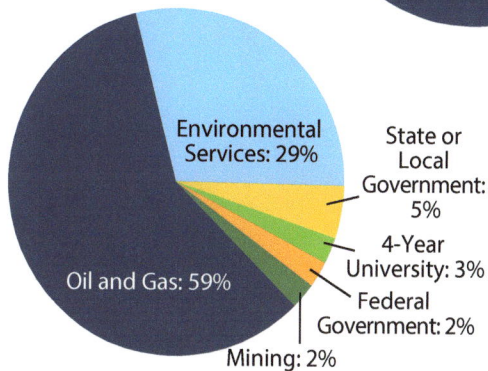

- Oil and Gas: 59%
- Environmental Services: 29%
- State or Local Government: 5%
- 4-Year University: 3%
- Federal Government: 2%
- Mining: 2%

Doctoral Graduates

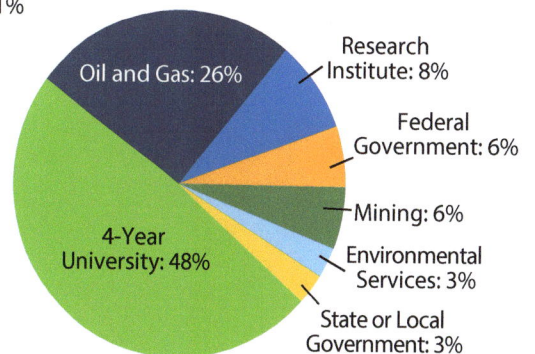

- Oil and Gas: 26%
- Research Institute: 8%
- Federal Government: 6%
- Mining: 6%
- Environmental Services: 3%
- State or Local Government: 3%
- 4-Year University: 48%

Starting salaries for graduates who accepted a job in the geosciences

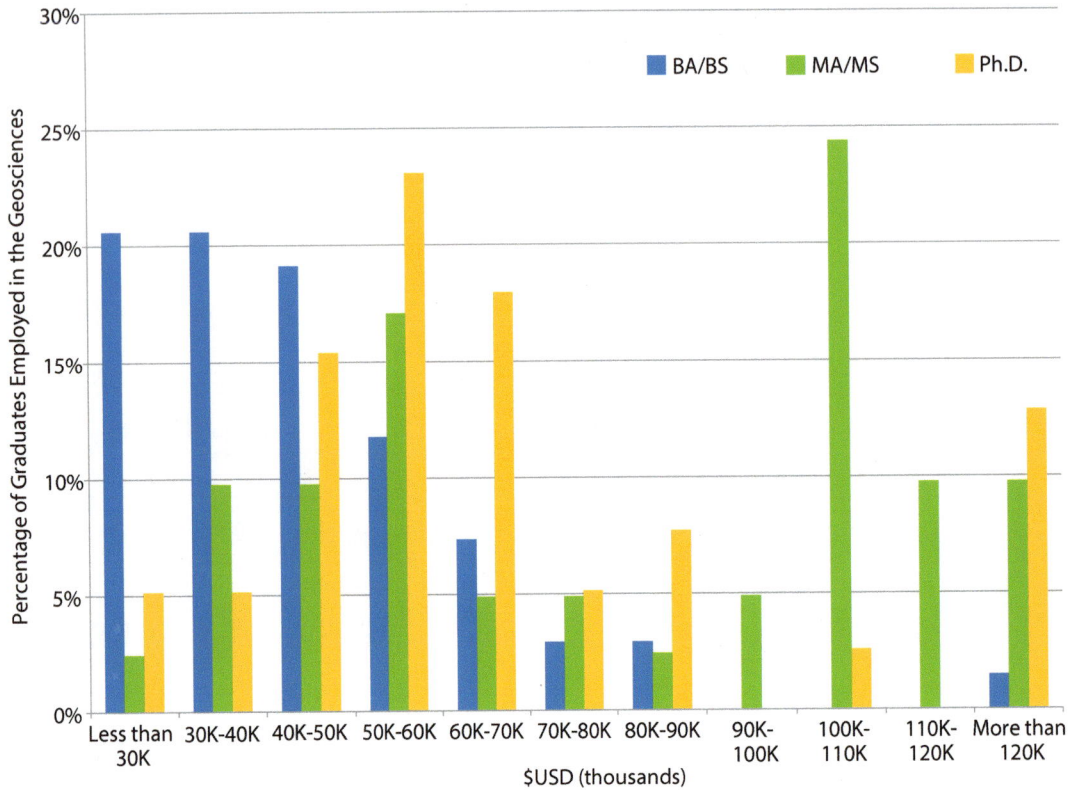

Additional compensation for graduates who accepted a job in the geosciences

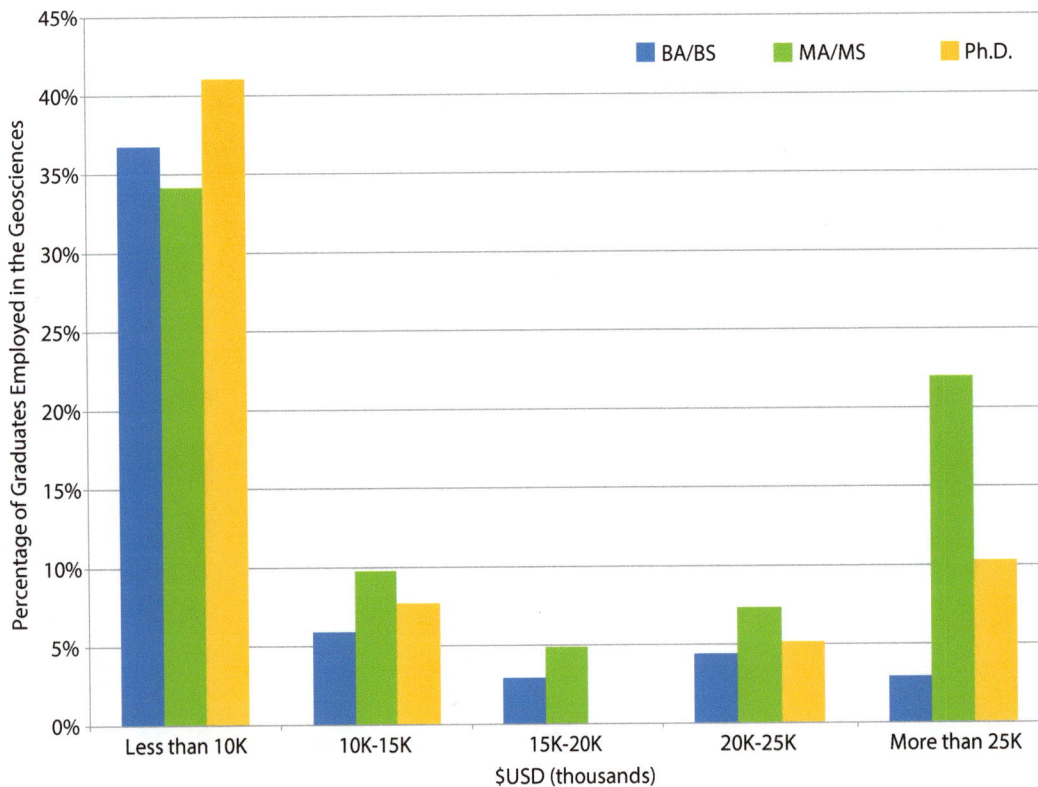

Resources identified as useful for finding geoscience jobs

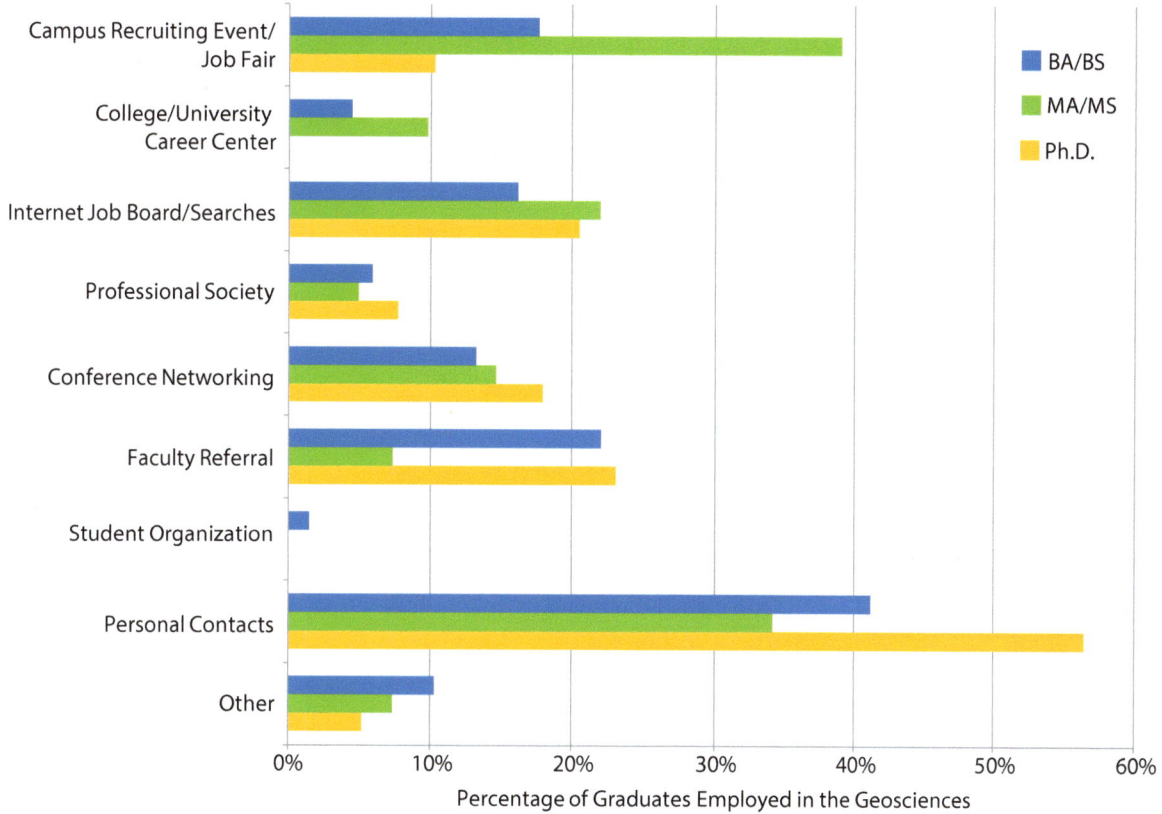

Percentage of Graduates Employed in the Geosciences

Other job opportunities offered to graduates who accepted a job in the geosciences

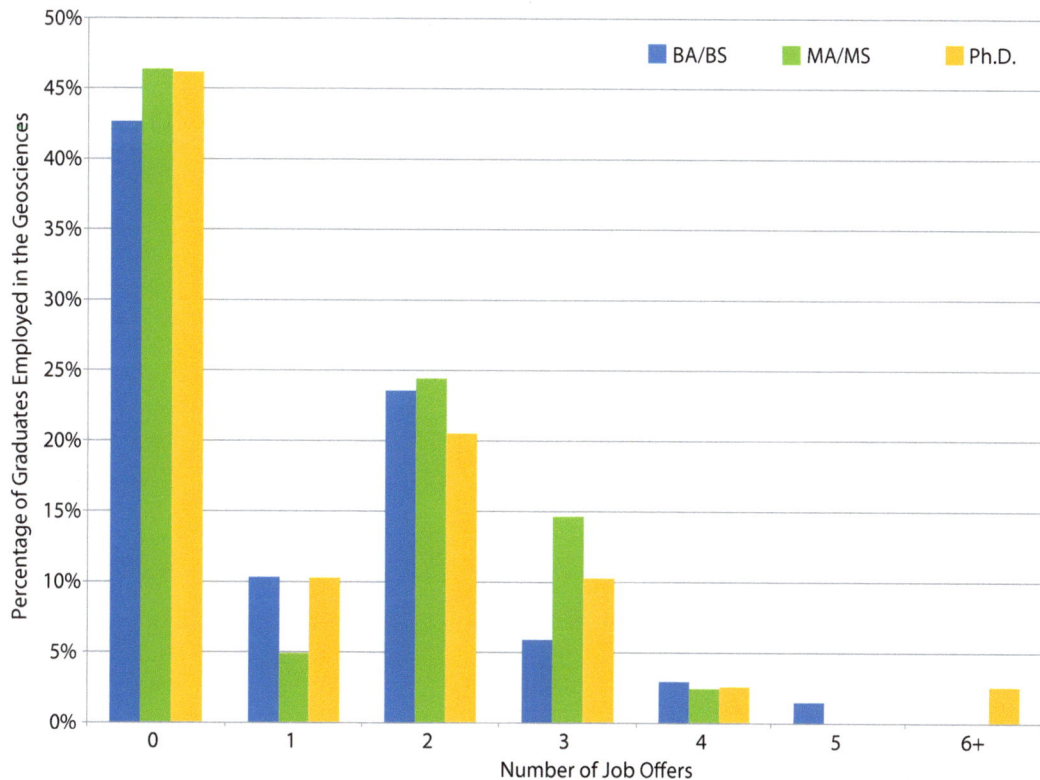

Number of Job Offers

States where graduates found employment in the geosciences

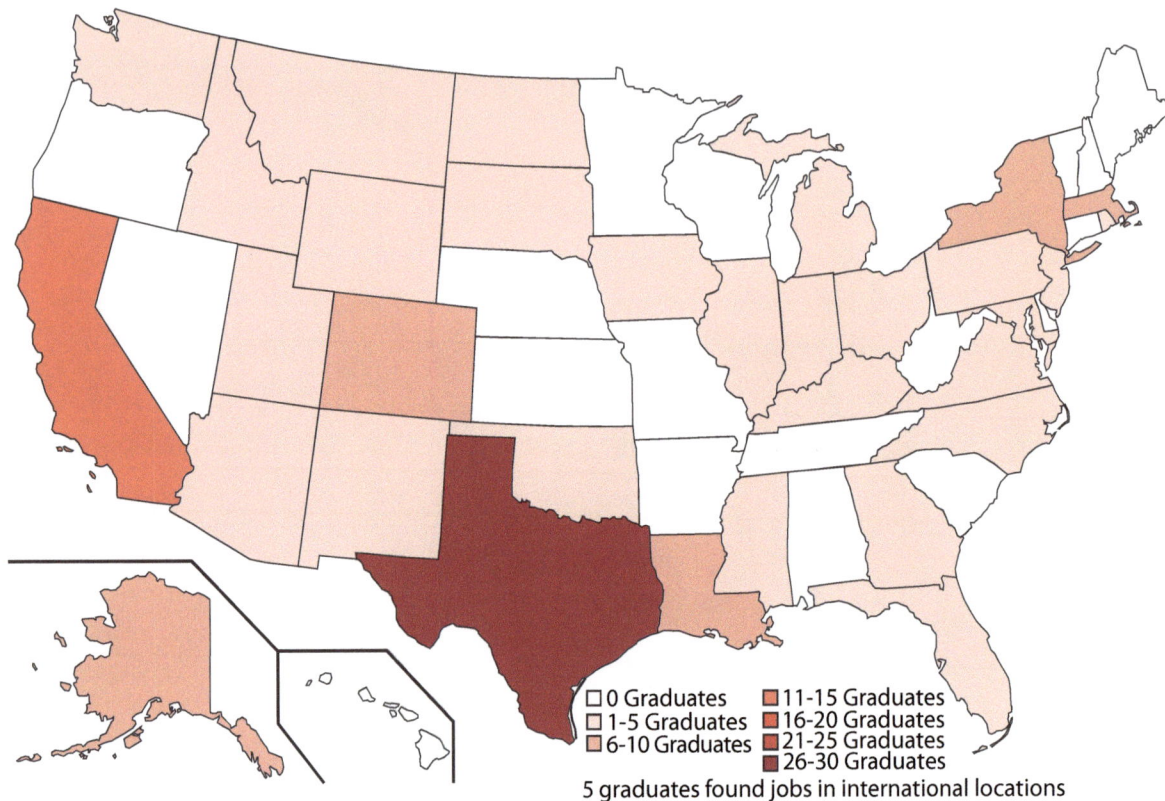

0 Graduates
1-5 Graduates
6-10 Graduates
26-30 Graduates

11-15 Graduates
16-20 Graduates
21-25 Graduates

5 graduates found jobs in international locations

Industries of interest for graduating students seeking a job within the geosciences

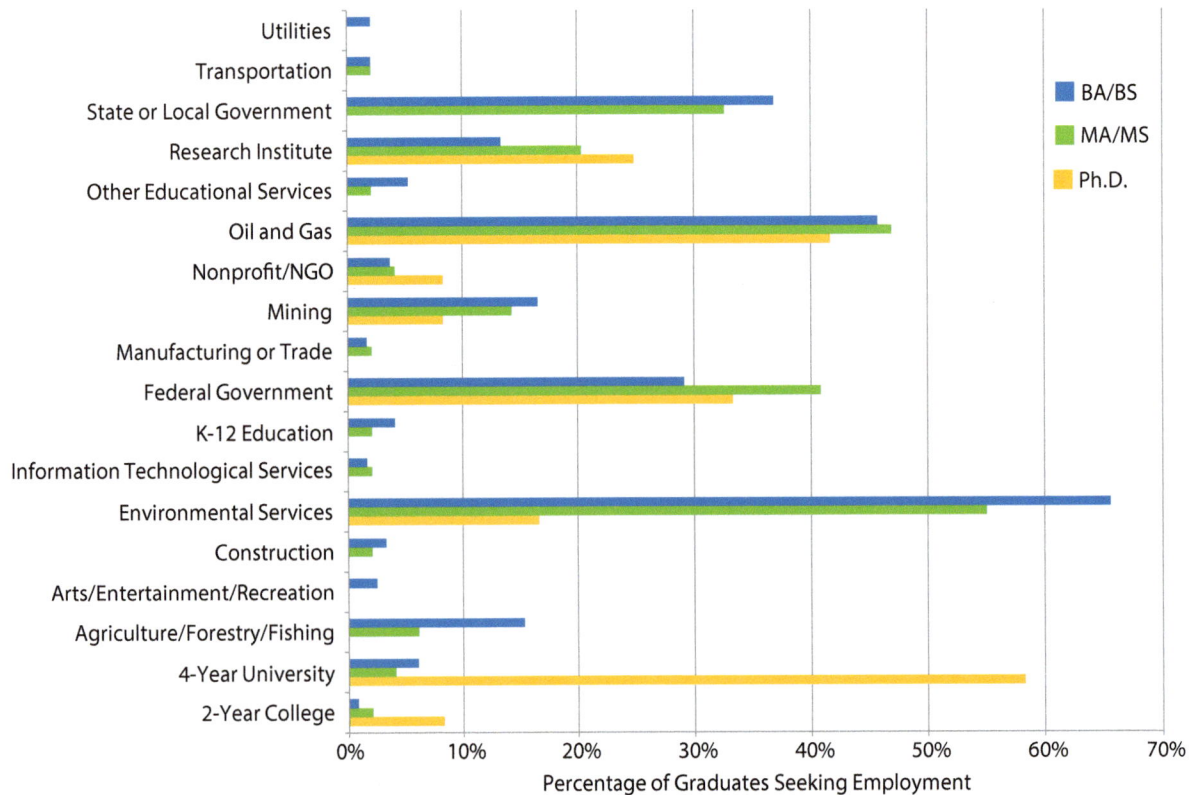

BA/BS
MA/MS
Ph.D.

Percentage of Graduates Seeking Employment

Industries of geoscience graduates' first jobs by degree field for the past two years***

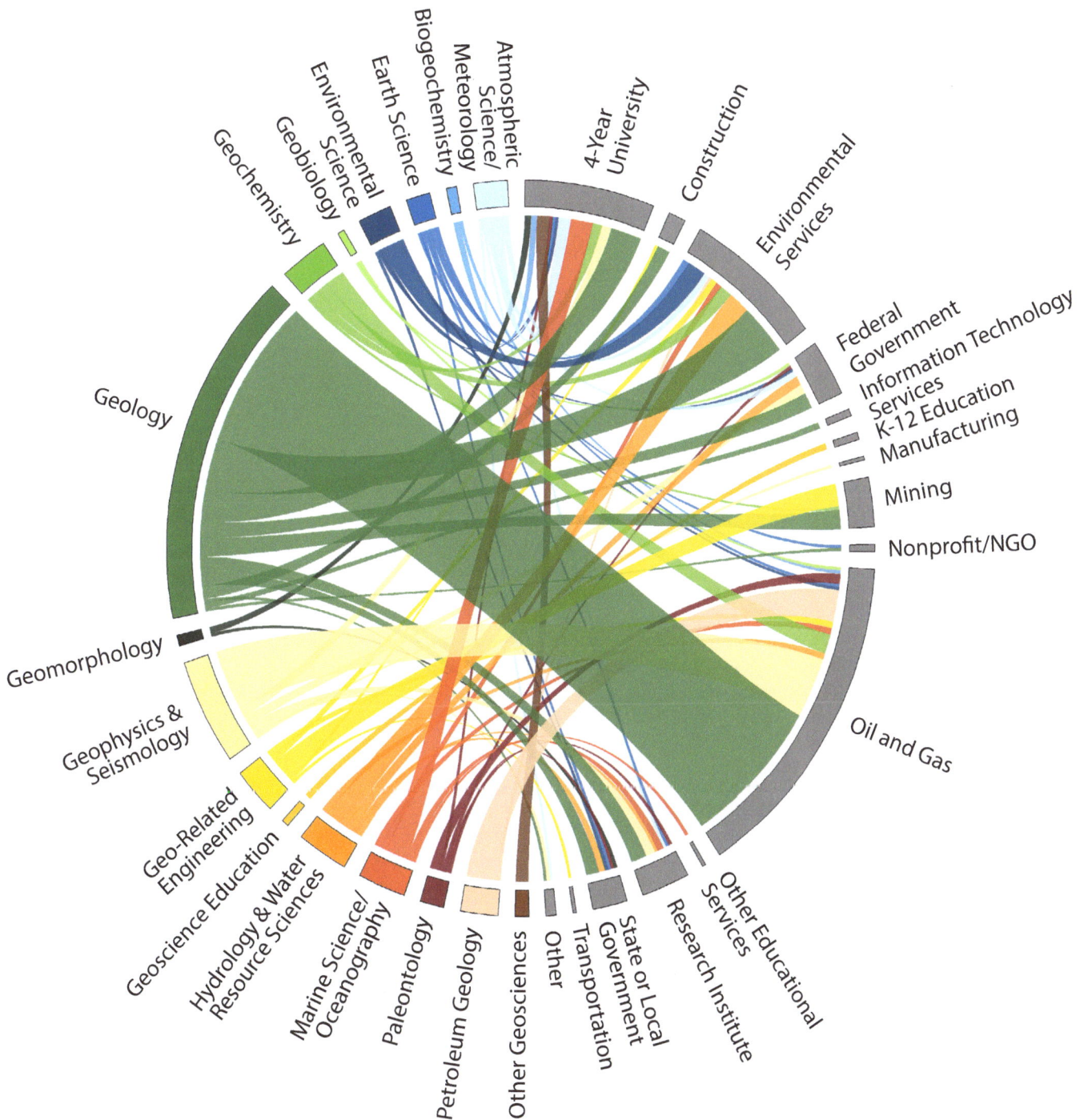

***The code for this visualization was modified from Kyzywinski, M. et al. Circos: an Information Aesthetic for Comparative Genomics. **Genome Res** (2009) 19:1693-1645

Future Plans: Working Outside of the Geosciences

Very few students are seeking or have secured jobs outside of the geosciences. Due to this, the figures displaying data about these graduates that either accepted or are seeking a job outside of the geosciences show the number of graduates regardless of degree level. Most of the graduates that have accepted a job position outside of the geosciences chose these positions because they wanted to pursue other interests, wanted a geoscience job but had trouble getting hired, and/or need to earn money to help pay for student loans or other life expenses. A handful of graduates commented on their desire for a science education job teaching the earth sciences or a job in other areas that allow the use of their geoscience knowledge. While these types of positions may not be considered traditional geoscience careers, AGI considers them still within the geosciences workforce.

Those graduates that had accepted a job outside of the geosciences were asked to provide more details about their jobs. The industries most often hiring students into jobs they felt were outside of the geosciences include four-year universities, the federal government, and other educational services outside of formal education. These graduates on average make quite a bit less in their starting salary that those graduates with jobs in the geosciences. However, most of these graduates with jobs outside of the geosciences found their job using similar resources as the graduates employed in the geosciences, such as through personal contacts and internet searches.

Graduating students seeking or have accepted a job position outside the geosciences

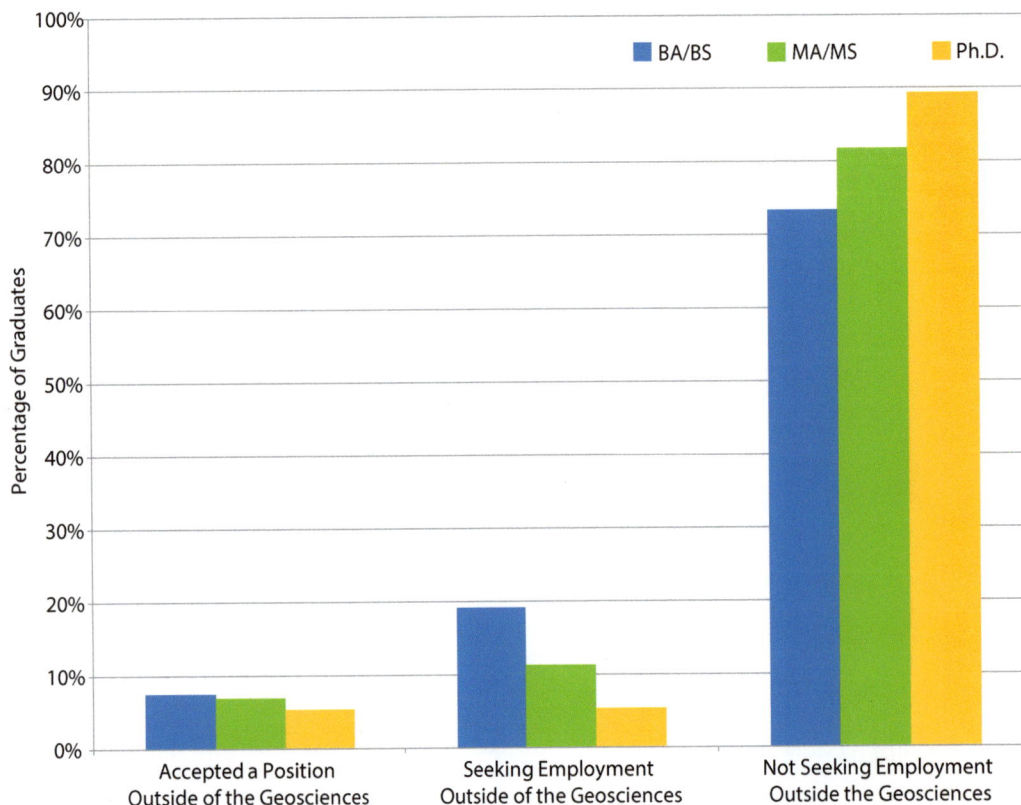

Industries where graduating students have accepted a job outside the geosciences

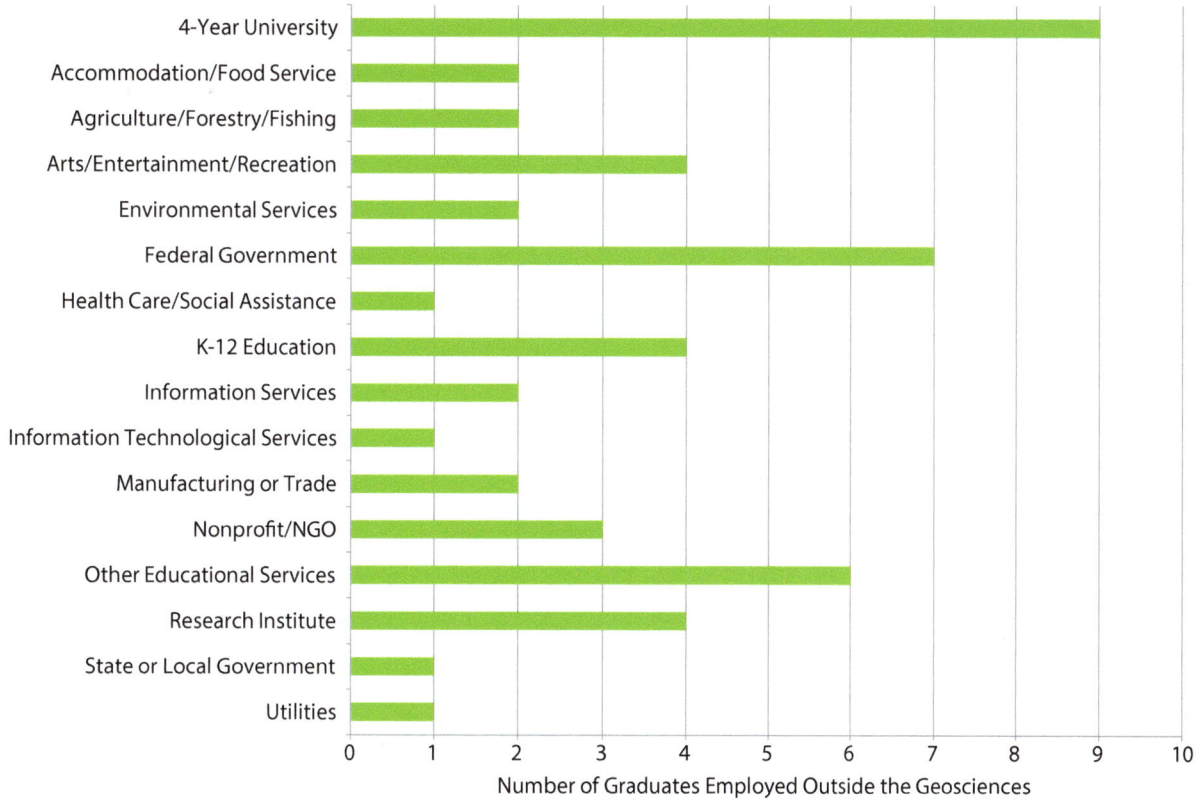

Number of Graduates Employed Outside the Geosciences

Starting salaries for graduating students that accepted a job outside the geosciences

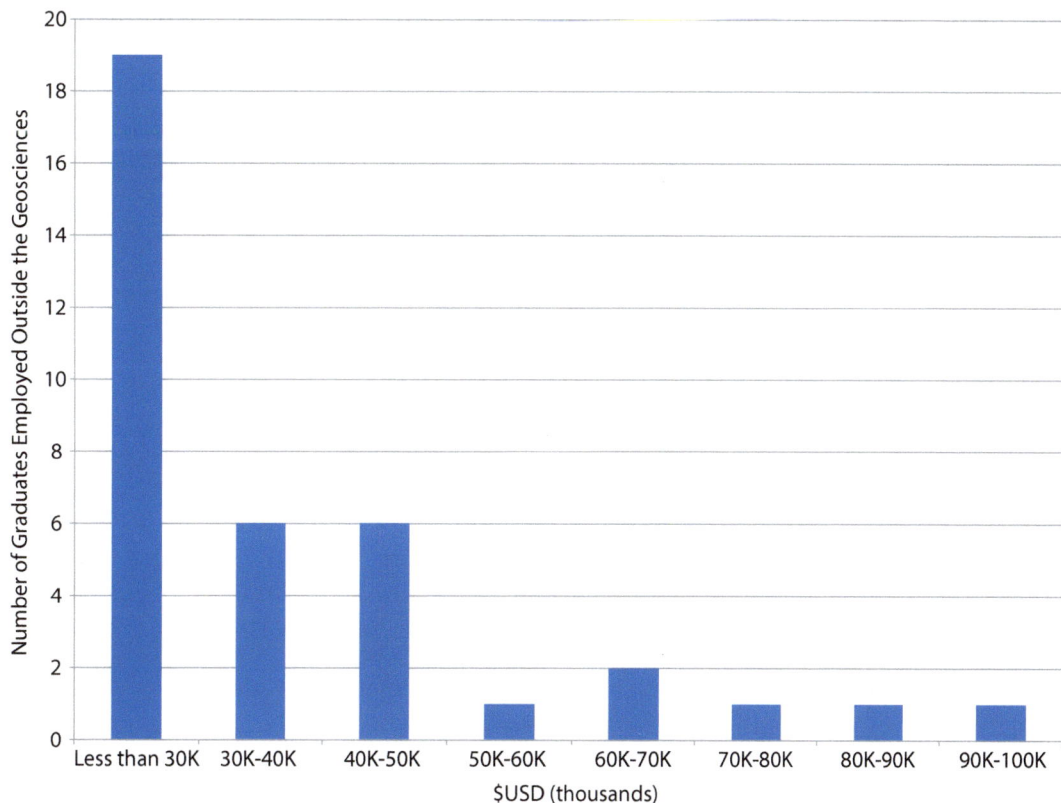

$USD (thousands)

Resources identified by graduating students as useful for finding non-geoscience jobs

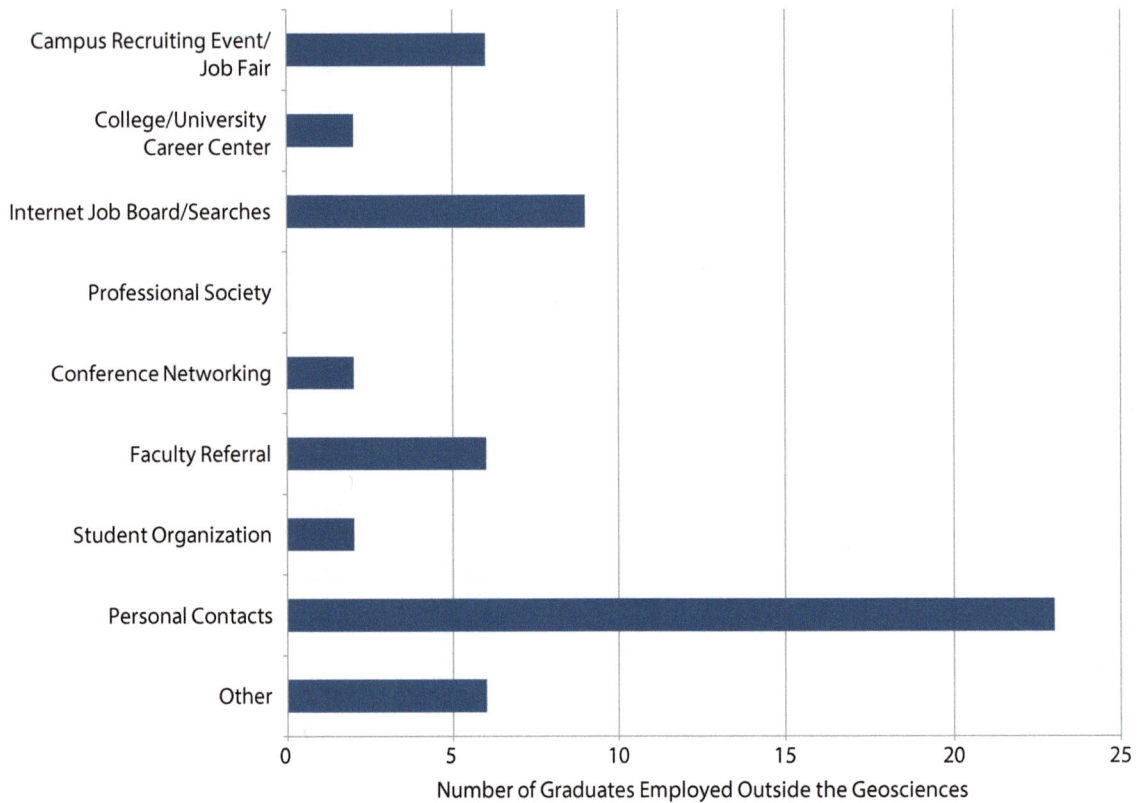

Photo by Artur Pacyga from the AGI 2014 Life in the Field photo contest.

Appendices

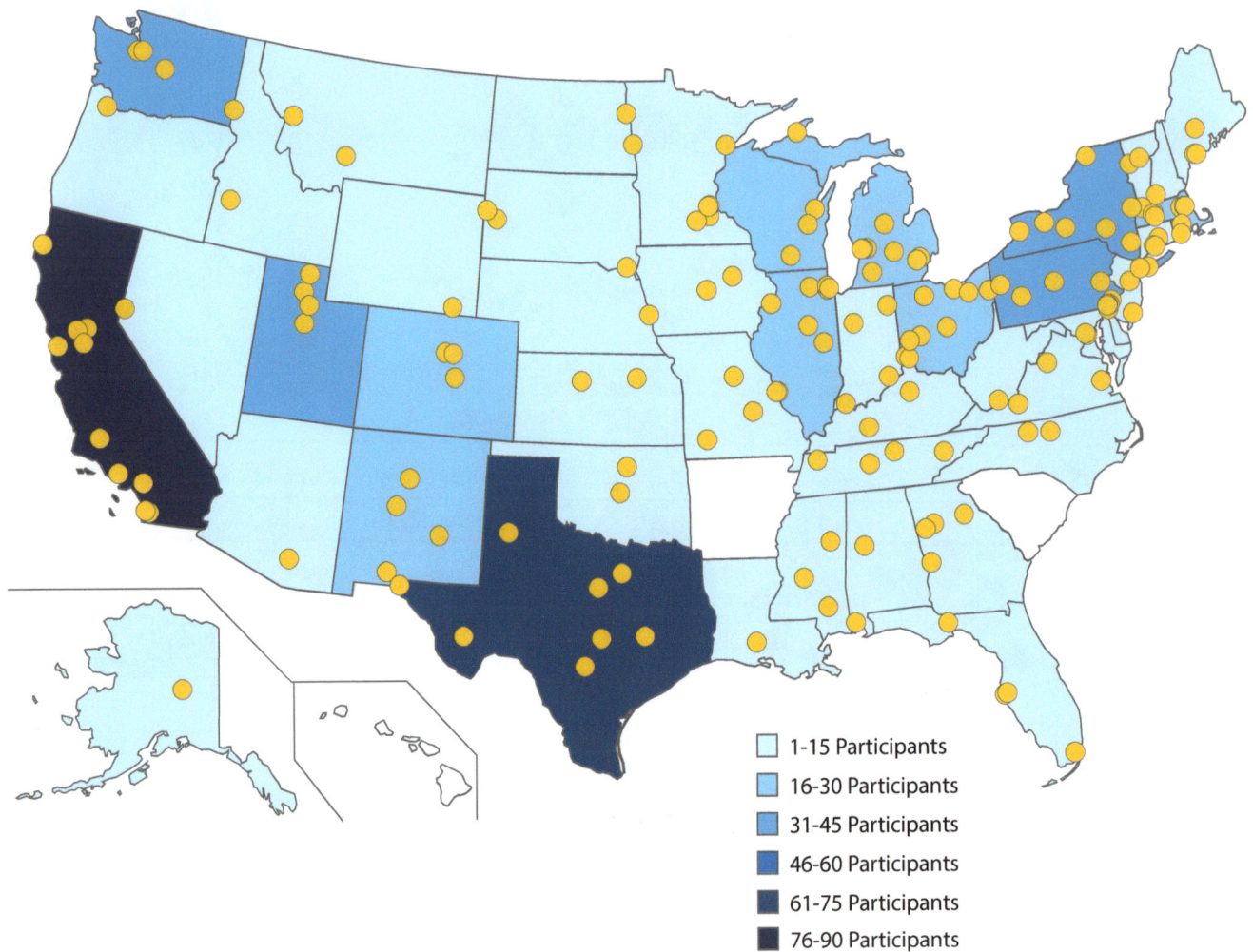

☐	1-15 Participants
☐	16-30 Participants
☐	31-45 Participants
☐	46-60 Participants
■	61-75 Participants
■	76-90 Participants

Appendix I

The following is a list of all the institutions and departments with graduating students that took AGI's Geoscience Exit Survey in the 2013-2014 academic year.

University, Department

Amherst College, Department of Geology

Augustana College, Department of Geology

Black Hills State University, Department of Environmental Physical Sciences

Boise State University, Department of Geosciences

Boston College, Department of Earth and Environmental Science

Bowdoin College, Department of Earth and Oceanographic Sciences

Bowling Green State University, Department of Geology

Brigham Young University, Department of Geological Sciences

Brown University, Department of Geological Sciences

Bryn Mawr College, Department of Geology

California State University-Bakersfield, Department of Geology

California State University-Fullerton, Department of Geological Sciences

California State University-Northridge, Department of Geological Sciences

California State University-Sacramento, Department of Geology

Calvin College, Department of Geology, Geography, and Environmental Studies

Carleton College, Department of Geology

Central Michigan University, Department of Earth and Atmospheric Sciences

Central Washington University, Department of Geological Sciences

Colby College, Department of Geology

College of William and Mary, Department of Geology

Colorado College, Department of Geology

Colorado School of Mines, Department of Geology and Geological Engineering

Columbus State University, Department of Earth and Space Sciences

Concord University, Department of Physical Sciences

Cornell University, Department of Earth and Atmospheric Sciences

Eastern Michigan University, Department of Geography and Geology

Eastern New Mexico University, Department of Physical Sciences

Eckerd College, Department of Geosciences

Eckerd College, Department of Marine Sciences

Florida State University, Department of Earth, Ocean, and Atmospheric Sciences

Fort Hays State University, Department of Geosciences

Georgia State University, Department of Geosciences

Grand Valley State University, Department of Geology

Guilford College, Department of Geology and Earth Sciences

Gustavus Adolphus College, Department of Geology

Hanover College, Department of Geology

Hope College, Department of Geological and Environmental Sciences

Humboldt State University, Department of Geology

Illinois State University, Department of Geology

Indiana University of Pennsylvania, Department of Geoscience

Indiana University-Purdue University Fort Wayne, Department of Geosciences

Iowa State University, Department of Geological and Atmospheric Sciences

James Madison University, Department of Geology and Environmental Sciences

Kansas State University, Department of Geology

Keene State College, Department of Geology

Kutztown University of Pennsylvania, Department of Physical Sciences

Metropolitan State University of Denver, Department of Earth and Environmental Sciences

Miami University of Ohio, Department of Geology and Environmental Earth Science

Michigan State University, Department of Geological Sciences

Michigan Technological University, Department of Geological/Mining Engineering and Sciences

Middle Tennessee State University, Department of Geosciences

Middlebury College, Department of Geology

Millsaps College, Department of Geology

Mississippi State University, Department of Geosciences

Missouri State University, Department of Geography, Geology, and Planning

Missouri University of Science and Technology, Department of Geosciences and Geological and Petroleum Engineering

MIT, Department of Earth, Atmospheric, and Planetary Sciences

Montana State University, Department of Geology

Mount Holyoke College, Department of Geology

New Mexico Institute of Mining and Technology, Department of Earth and Environmental Sciences

New Mexico State University, Department of Geological Sciences

North Dakota State University, Department of Geosciences

Northern Illinois University, Department of Geology and Environmental Geosciences

Northwestern University, Department of Earth and Planetary Sciences

Norwich University, Department of Geology

Oberlin College, Department of Geology

Ohio State University, School of Earth Sciences

Oklahoma State University, Department of Geology

Pace University, Environmental Sciences Program

Pacific Lutheran University, Department of Geoscience

Pennsylvania State University, Department of Geosciences

Portland State University, Department of Geology

Purdue University, Department of Earth and Atmospheric Sciences

Queens College-CUNY, School of Earth and Environmental Sciences

Rensselaer Polytechnic Institute, Department of Earth and Environmental Sciences

Richard Stockton College of New Jersey, Department of Geology

Rutgers University, Department of Earth and Planetary Sciences

San Diego State University, Department of Geological Sciences

Slippery Rock University, Department of Geography, Geology, and the Environment

Smith College, Department of Geosciences

South Dakota School of Mines and Technology, Department of Geology and Geological Engineering

St. Lawrence University, Department of Geology

St. Louis University, Department of Earth and Atmospheric Sciences

St. Norbert College, Department of Geology

Sul Ross State University, Department of Biology, Geology, and Physical Sciences

SUNY Geneseo, Department of Geological Sciences

SUNY College at Oneonta, Department of Earth and Atmospheric Sciences

Tarleton State University, Department of Geoscience

Temple University, Department of Earth and Environmental Science

Tennessee Tech University, Department of Earth Sciences

Texas A&M University, School of Geosciences

Trinity University, Department of Geosciences

Tufts University, Department of Earth and Ocean Sciences

University at Buffalo, Department of Geology

University of Akron, Department of Geosciences

University of Alabama, Department of Geological Sciences

University of Alaska-Fairbanks, Department of Geology and Geophysics

University of Arizona, Department of Geosciences

University of California-Berkeley, Department of Earth and Planetary Science

University of California-Davis, Department of Earth and Planetary Sciences

University of California-San Diego, Scripps Institution of Oceanography

University of Cincinnati, Department of Geology

University of Delaware, Department of Geological Sciences

University of Georgia, Department of Geology

University of Idaho, Department of Geological Sciences

University of Illinois at Chicago, Department of Earth and Environmental Sciences

University of Illinois, Department of Geology

University of Kentucky, Department of Earth and Environmental Sciences

University of Louisiana at Lafayette, Department of Geology

University of Maryland, Department of Geology

University of Maryland-Baltimore County, Department of Environmental Engineering

University of Miami, Department of Geological Sciences

University of Michigan, Department of Earth and Environmental Sciences

University of Minnesota-Duluth, Department of Geosciences

University of Minnesota-Twin Cities, Department of Earth Sciences

University of Missouri, Department of Soil, Environmental, and Atmospheric Science

University of Montana, Department of Geosciences

University of Nebraska-Omaha, Department of Geography/Geology

University of Nevada-Reno, Mackay School of Earth Science and Engineering

University of New Haven, Department of Environmental Science

University of New Mexico, Department of Earth and Planetary Sciences

University of North Carolina at Chapel Hill Department of Geosciences

University of North Dakota, School of Geology and Geological Engineering

University of Northern Iowa, Department of Earth Science

University of Oklahoma, School of Geology and Geophysics

University of Pennsylvania, Department of Earth and Environmental Science

University of Rhode Island, Department of Geosciences

University of South Alabama, Department of Earth Sciences

University of South Dakota, Department of Earth Sciences

University of South Florida, Department of Geology

University of Southern Indiana, Department of Geology and Physics

University of Southern Maine, Department of Geosciences

University of Southern Mississippi, Department of Marine Science

University of St. Thomas, Department of Geology

University of Tennessee at Knoxville, Department of Earth and Planetary Sciences

University of Tennessee at Martin, Department of Agriculture and Applied Sciences

University of Texas at Arlington, Department of Earth and Environmental Sciences

University of Texas at Austin, Jackson School of Geosciences

University of Texas at El Paso, Department of Geological Sciences

University of the Pacific, Department of Earth and Environmental Sciences

University of Utah, College of Mines and Earth Sciences

University of Washington, Department of Earth and Space Sciences

University of Washington, Department of Oceanography

University of West Georgia, Department of Geosciences

University of Wisconsin-Madison, Department of Geology and Geophysics

University of Wisconsin-Oshkosh, Department of Geology

University of Wyoming, Department of Geology and Geophysics

Utah State University, Department of Geology

Vassar College, Department of Earth Sciences

Virginia Polytechnic Institute and State University, Department of Geosciences

Washington University in St. Louis, Department of Earth and Planetary Sciences

Weber State University, Department of Geosciences

Wesleyan University, Department of Earth and Environmental Sciences

West Chester University, Department of Geology and Astronomy

Western Kentucky University, Department of Geography and Geology

Western Michigan University, Department of Geosciences

Wheaton College, Department of Geology and Environmental Science

Williams College, Department of Geosciences

Wittenberg University, Department of Geology

Wright State University, Department of Earth and Environmental Sciences

Yale University, Department of Geology and Geophysics

Youngstown State University, Department of Geological and Environmental Sciences

Appendix II

Carnegie Classifications of Institutions of Higher Learning (http://classifications.carnegiefoundation.org/)

This classification system was used for some of the analysis of the Spring 2013 results of AGI's Geoscience Student Exit Survey. The following are the definitions for the classification system and the participating institutions belonging to each category as defined and categorized by the Carnegie Foundation for the Advancement of Teaching.

Baccalaureate Colleges — Arts & Sciences (Bac/A&S)

Baccalaureate Colleges — Diverse Fields (Bac/Diverse)

Includes institutions where baccalaureate degrees represent at least 10 percent of all undergraduate degrees and where fewer than 50 master's degrees or 20 doctoral degrees were awarded during the update year. Excludes Special Focus Institutions and Tribal Colleges.

Among Institutions where bachelor's degrees represented at least half of all undergraduate degrees, those with at least half of bachelor's degree majors in arts and science fields were included in the "Arts & Sciences" group, while the remaining institutions were included in the "Diverse Fields" group.

Exit Survey Departments (Bac/A&S):
Amherst College
Augustana College
Bowdoin College
Bryn Mawr College
Calvin College
Carleton College
Colby College
Colorado College
Eckerd College
Guilford College
Gustavus Adolphus College
Hanover College
Hope College
Middlebury College
Millsaps College
Mount Holyoke College
Oberlin College
Smith College
St. Lawrence University

St. Norbert College
Vassar College
Wesleyan University
Wheaton College
Williams College
Wittenberg University

Exit Survey Departments (Bac/Diverse)*
Concord University
Metropolitan State University of Denver

Master's Colleges and Universities — Larger Programs (Master's/L)

Master's Colleges and Universities — Medium Programs (Master's/M)

Master's Colleges and Universities — Smaller Programs (Master's/S)

Generally includes institutions that awarded at least 50 master's degrees and fewer than 20 doctoral degrees during the update year (with occasional exceptions). Excludes Special Focus Institutions and Tribal Colleges.

Master's program size was based on the number of master's degrees awarded during the update year. Those awarding at least 200 degrees were included among larger programs; those awarding 100-199 were included among the medium programs; and those awarding 50-99 were included among the smaller programs. The smaller programs group also includes institutions that awarded fewer than 50 master's degrees if (a) their Enrollment Profile classification is Exclusively Graduate/Professional or (b) their Enrollment Profile classification is Majority Graduate/Professional and they awarded more graduate/professional degrees than undergraduate degrees.

Exit Survey Departments (Master's/L):
Boise State University
California State University-Bakersfield
California State University-Fullerton
California State University-Northridge
California State University-Sacramento
Columbus State University
Eastern Michigan University
Fort Hays State University
Grand Valley State University
Indiana University-Purdue University
James Madison University

Kutztown University of Pennsylvania
Missouri State University
Norwich University
Queens College-CUNY
Slippery Rock University
Sul Ross State University
Tarleton State University
Tennessee Tech University
University of New Haven
University of Northern Iowa
University of Southern Indiana
University of Southern Maine
University of West Georgia
University of Wisconsin-Oshkosh
West Chester University
Western Kentucky University
Youngstown State University

Exit Survey Departments (Master's/M):

Central Washington University
Humboldt State University
New Mexico Institute of Mining and Technology
Pacific Lutheran University
Richard Stockton College of New Jersey
Trinity University
University of Minnesota-Duluth
University of Tennessee at Martin
Weber State University

Exit Survey Departments (Master's/S):

Black Hills State University
Eastern New Mexico State University
Keene State College
SUNY Geneseo
SUNY College at Oneonta

Research Universities — Very High Research Activity (RU/VH)

Research Universities — High Research Activity (RU/H)

Doctoral/Research Universities (DRU)

Includes institutions that awarded at least 20 research doctoral degrees during the update year (excluding doctoral-level degrees that qualify recipients for entry into professional practice, such as the JD, MD, PharmD, DPT, etc.). Excludes Special Focus Institutions and Tribal Colleges.

Doctorate-granting institutions were assigned to one of three categories based on a measure of research activity. For more information about the analysis of research activity, please visit http://classifications.carnegiefoundation.org/methodology/basic.php.

Exit Survey Departments (RU/VH):

Brown University
Cornell University
Florida State University
Georgia State University
Iowa State University
Michigan State University
Mississippi State University
MIT
Montana State University
North Dakota State University
Northwestern University
Ohio State University
Pennsylvania State University
Purdue University
Rensselaer Polytechnic Institute
Rutgers University
Texas A&M University
Tufts University
University of Arizona
University at Buffalo
University of California-Berkeley
University of California-Davis
University of California-San Diego
University of Cincinnati
University of Delaware
University of Georgia
University of Illinois at Chicago
University of Illinois
University of Kentucky
University of Maryland
University of Miami
University of Michigan
University of Minnesota-Twin Cities
University of Missouri
University of New Mexico
University of North Carolina at Chapel Hill
University of Oklahoma
University of Pennsylvania
University of South Florida
University of Tennessee at Knoxville
University of Texas at Austin
University of Utah
University of Washington
University of Wisconsin-Madison
Virginia Polytechnic Institute and State University
Washington University in St. Louis
Yale University

Exit Survey Departments (RU/H):

Boston College
Bowling Green State University
Brigham Young University
College of William and Mary
Colorado School of Mines
Kansas State University
Miami University of Ohio
Michigan Technological University
Missouri University of Science and Technology
New Mexico State University
Northern Illinois University
Oklahoma State University
Portland State University
San Diego State University
St. Louis University
Temple University
Texas Tech University
University of Akron
University of Alabama
University of Alaska-Fairbanks
University of Idaho
University of Louisiana at Lafayette
University of Maryland-Baltimore County
University of Montana
University of Nevada-Reno
University of North Dakota
University of Rhode Island
University of South Alabama
University of South Dakota
University of Southern Mississippi
University of Texas at Arlington
University of Texas at El Paso
University of Wyoming
Utah State University
Western Michigan University
Wright State University

Exit Survey Departments (DRU):

Central Michigan University
Illinois State University
Indiana University of Pennsylvania
Middle Tennessee State University
Pace University
University of Nebraska-Omaha
University of St. Thomas
University of the Pacific

Special Focus Institutions — Schools of Engineering (Spec/Engg)

Institutions awarding baccalaureate or higher-level degrees where a high concentration of degrees (above 75%) is in a single field or set of related fields. Excludes Tribal Colleges.

Exit Survey Departments (Spec/Engg)*:

South Dakota School of Mines and Technology

*Institutions in this classification where not included in comparisons using the Carnegie Classification system due to the small number of institutions in the Exit Survey belonging to the particular classification.

www.ingramcontent.com/pod-product-compliance
Lightning Source LLC
Chambersburg PA
CBHW052044190326
41520CB00002BA/176